'AI appears to be eating the world. How then to understand where this is taking us, both the promises and the threats? This book is for you. A down-to-earth tour that will help you understand how the algorithms are increasingly taking on human roles.'
Toby Walsh, Chief Scientist at the UNSW AI Institute and author of *Faking It* and *Machines Behaving Badly*

'AI and algorithms are reshaping our world at an unprecedented pace, with buzzwords like neural networks, deep learning and transformers dominating the headlines. But what significance do these terms hold for you? Professor Kowalkiewicz's eagerly awaited book provides clear and accessible answers to this question.'
Chong Wang, AI research scientist, Apple

'A very accessible book for anyone interested in the long-term opportunities for businesses, individuals and society as technology penetrates ever more corners of the economy and our lives ... enlightening, informative and encouraging.'
Simon Dale, Vice-President, Adobe

'A damn well-written book ... a thrilling, entertaining whirlwind tour of the different AI algorithms and their industry applications ... This is the book I will, from now on, recommend to anyone who wants to understand the implications that a new generation of AI-powered algorithms will have on human society and how to best work together with them.'
Tobias Lange, Senior Vice President, Siemens

'This book is right on time for us to understand more of the good, the bad and the ugly of algorithms. An eye-opener about how ubiquitous digital helpers and algorithms have become in our daily lives.'
Mario Herger, author of Gamification at Work

'This is a fantastic and timely book about the opportunities and risks of algorithms ... It concludes with a series of much-needed recommendations that provide guidance on how we can all learn to live and prosper with our digital minions.'
Jonathan Reichental, author of Smart Cities for Dummies

'Kowalkiewicz deftly charts the so-called "march of the algorithms" and how they have come to dominate so much of our world ... This is an optimistic guide to tilting our modern economy of algorithms toward opportunity instead of disaster.'
Rohit Bhargava, author of Non-Obvious Megatrends

'From automating mundane tasks to challenging ethical boundaries, this book delves into the heart of our algorithm-driven future, offering insightful rules for thriving in this new era.'
Aleksandra Przegalińska, Vice-Rector for International Cooperation and ESR, Kozminski University and Head of the HumanRace Research Center

THE ECONOMY OF ALGORITHMS

AI and the Rise of the Digital Minions

Marek Kowalkiewicz

BRISTOL
UNIVERSITY
PRESS

Published by arrangement with La Trobe University Press,
an imprint of Schwartz Books Pty Ltd

First published in Great Britain in 2024 by

Bristol University Press
University of Bristol
1-9 Old Park Hill
Bristol
BS2 8BB
UK
t: +44 (0)117 374 6645
e: bup-info@bristol.ac.uk

Details of international sales and distribution partners are available at
bristoluniversitypress.co.uk

© Marek Kowalkiewicz, 2024

British Library Cataloguing in Publication Data
A catalogue record for this book is available from the British Library

ISBN 978-1-5292-4246-1 paperback
ISBN 978-1-5292-4247-8 ePub
ISBN 978-1-5292-4248-5 ePdf

Cover design: Mecob

Bristol University Press uses environmentally responsible
print partners.

Printed and bound in Great Britain by CPI Group (UK) Ltd,
Croydon, CR0 4YY

FSC
www.fsc.org
MIX
Paper | Supporting
responsible forestry
FSC® C013604

To Anetta, Filip and Zofia,
my constants

Contents

Acknowledgements   viii

Introduction: The Economy of Algorithms Is Here   1

Part I   How Did We Get Here?
1   The March of the Algorithms   9
2   New Agents Enter the Economy   31
3   New Digital Frontiers   51

Part II   How Algorithms Affect Us All
4   How Do You Advertise to a Fridge?   71
5   Can an Algorithm Take Your Job?   89
6   Do Algorithms Dream of Electric Sheep?   110

Part III Nine Rules for the Age of Algorithms
7   Be the Minion Master: Revenue Automation   137
8   Be Relentlessly Curious: Continuous Evolution   155
9   Be Boldly Optimistic: Relationship Saturation   176

Conclusion: Human Agency   190
Notes   193

Acknowledgements

This book was almost six years in the making. It was back in 2018, when I spoke about the economy of algorithms at the World Science Festival, that the idea of writing it was born. My session was in a large tent, not dissimilar from those you see in Germany during Oktoberfest. People were having beers and snacks, and I could see they enjoyed hearing about algorithmic agents. I needed to write a book about digital minions. But how do you go about doing that? I had never written a book for a general audience before.

Somehow, I thought writing it would be a solitary task. In my mind, I would sit down one day and start writing – and then repeat this every day until I finished the book, caught in a daily grind like Taleb's turkey. But it didn't work. I kept going in circles, writing and rewriting chapters. The content wasn't working, and the structure wasn't coming together. I knew how to write a couple of pages of engaging content – but writing a couple of hundred pages that worked well? I struggled.

Trying to break the cycle, I began studying how successful authors worked. I was using every opportunity to learn. I remember bumping into an author I admired on a cold wintery evening in 2019. Rachel Botsman, whose book I had just finished, boarded the same tiny shuttle bus I was on. We were both attending the World Economic Forum in Davos. She must have thought I was out of my mind, as I kept asking about her writing process instead of the topics the forum was focusing on that year. I learned a lesson from our conversation: if you want to write, you have to make time for it. But stalking authors on shuttle buses isn't a very efficient way to learn the tools of the trade, I thought. There had to be an easier way.

Thankfully, almost every author happily shares their process of book-writing at the back of the book itself. The acknowledgments section often gives away what worked for its author and sometimes what didn't. So I started reading the final sections of other books, in some cases leaving the rest of the book for later.

That's when I realised that it's not just the process that matters in writing. That part is relatively simple: most authors do sit down and write daily – though it's much harder than it sounds! It's also people. Even though only one or two names typically end up on the cover, great books have a much larger team behind them. And, for me, everything changed when I realised that. I began reaching out.

When the idea for my book was still emerging, Veny Armanno, at Queensland Writers Centre, taught me how to shift from the micro level (stories that are just a few pages long) to the macro (the whole book). Paula Dootson was patient enough to read several versions of my earliest chapters, and she provided ample encouragement, even though – let's face it – the early versions were terrible. Michael Rosemann influenced many ideas you read about in this book – in particular, his concept of the 'economy of things' inspired my realisation that algorithms can become economic agents. He also developed the first set of innovation lenses, mentioned in Chapter 1.

In 2022, I took a few months' sabbatical leave to focus solely on this book. Within days, I contacted Kate Eltham, who profoundly understands the book industry. She was generous enough to spend time explaining how writing and publishing works. She patiently walked me through all the basics, including the various publishing strategies authors can use. She even helped me set my priorities and expectations.

Over the next few months, Kath Walters taught me how to write a book without getting stuck. Her approach is similar to that which software developers use when writing computer code, so it felt immediately familiar. To all the geeks reading this section: I followed Agile principles. Sprints, daily scrums, user stories, kanban boards and even pomodoro timers: I used them all.

And then there was my 'writing accountability group'. Jane Langof, Paul Atkins, Rob Crowe and I spent countless hours in our group, meeting regularly to sit down and write. Each of us showed up in a virtual Zoom room, joining from a different city in Australia. During Covid-19 lockdowns, this was the only way we could do it. One of us –I won't disclose who – would occasionally fall asleep while writing. I still chuckle when I think about seeing them nod off on my screen. We haven't had a chance to meet each other in person yet. I am sure one day we will.

Toby Walsh's books showed me that it is possible to introduce academic topics engagingly. I had met Toby just two days before I met Rachel Botsman. Not in Davos but in Munich, at a conference called DLD (Digital–Life–Design), where I get to replenish my ideation reserves every year. The community at DLD is second to none. When, a year later, I told Toby that I was working on a book, he introduced me to his agent, Margaret Gee. Marg agreed to read my initial chapters and soon decided to represent me. Only later did I learn how lucky I was: Marg is among the best of the best in Australia.

I am grateful to everyone who read the early drafts of the book and gave me feedback, as well as those who discussed the concepts with me while I was writing it. Mario Herger is the fastest reader I know, and he even corrected some of the endnotes. Simon Dale and Tobias Lange were able to highlight what makes this book different from others, which helped me in the final round of edits. Peter Townson always helped me find better ways to frame and express my ideas. Other colleagues at the Queensland University of Technology helped me shape the ideas further: Kevin Desouza, Erwin Fielt, Nadine Ostern, Chelsea Phillips, Malmi Amadoru and Wasana Bandara. I've also been lucky to work with several Technical University of Munich colleagues on ideas that formed the book: Jörg Weking, Michael Weber, Maria Stöcker-Stafflinger and others. Many others might not even be aware that simply having a chat with me helped in writing this book: Ivano Bongiovanni, Friedrich Chasin (who braved the heat with me when visiting Watkins Steel), Dayna Williamson (who opted not to have 'toilet' on her business card), Simon McKee, Monte Walker, Stefan Hajkowicz, Gemma Alker and Jodie Pattinson. Paul Kurchina kept asking me, 'When's the book coming out?', providing the right amount of pressure and friendly support.

Importantly, this book wouldn't have come into existence if not for the ongoing support of my family. My wife, Anetta, showed fantastic patience every time I descended into the 'man cave' (our term for a tiny, windowless space under our house that once stored garden tools but became my writing corner) and stayed there for hours at a time. My kids, Filip and Zofia, have always cheered me on, though they still give me stern looks for starting the book with the 'f-word'. In more ways than one, I wrote this book for them, and I cannot fully express how thankful I am for their patience and support.

Lastly, this book wouldn't be here if not for my parents. Back in 1985, when I was a small kid in Cold War–era Poland, they somehow managed to buy me and my brother our first computer, an Atari 800XL. I still remember the first time we turned it on. It was the first time an algorithm did something I told it to.

Five letters and a block cursor appeared in the upper left corner of the screen:

READY
█

I typed:

```
10 PRINT "HELLO WORLD"
RUN
```

And my first digital minion responded:

HELLO WORLD

Introduction:
The Economy of
Algorithms Is Here

'Fuck the algorithm!'

In the heart of London, right behind Westminster Abbey, and just as close to the famous Big Ben, there is a short, narrow street called Great Smith Street. It runs parallel to the Thames. And, just like the river, the street has seen its share of the city's history – including the history of information technology. The world's first computer programmer, Ada Lovelace, was born just a few minutes away in December 1815. And one of the founders of modern computer science, Alan Turing, known for breaking Enigma ciphers in World War II, began his life a couple of miles away, a century later, in June 1912. Lovelace and Turing saw beauty in algorithms and their potential to make the world a better place. Quite possibly, they each used to stroll along the quiet street while pondering their work.

But on Sunday, 16 August 2020, the street was far from quiet. It was louder and angrier than Lovelace and Turing would ever have experienced it.

'Fuck the algorithm!' The chants intensified.

Hundreds of students were gathered in front of the Department for Education at 20 Great Smith Street, voicing their frustration. The students were the victims of an algorithm that had been allowed to decide their future, and they were not having it.

What is an *algorithm?* Essentially, an algorithm is a step-by-step procedure for solving a problem or performing a computation. If I were to avoid formal definitions here, I would say that an algorithm is like a recipe that a computer follows. Computers are useless without algorithms. They calculate, predict, optimise and do all sorts of other things that make our human lives easier. Some of them are very simple, others so complex that people cannot comprehend them.

'Fuck the algorithm!'

Ah, right. Let's get back to the Great Smith Street commotion.

In 2020, students in the United Kingdom couldn't sit their final-year exams due to the pandemic. These exams are critical for high-school leavers: university offers are conditional on the results. If a student's grade is too low, they won't get into their preferred university.

To get around this issue in England, the Office of Qualifications and Examinations Regulation (Ofqual) introduced an algorithm to calculate the missing exam grades. First, teacher-predicted grades for each student, ranking them within their school, were fed into the algorithm. Then, the historical grade distributions of schools from three prior years were entered. Finally, previous exam results, per subject, were added. Well, that's how the algorithm worked for students at larger schools. For smaller schools, if fifteen or fewer students were enrolled in a particular subject, only teacher predictions were considered.

What resulted sparked the student protests. Students at smaller – typically private – schools, benefited from teacher optimism, or so-called 'grade inflation': the algorithm reflected teachers' predictions and scored them high. But everyone else fell into a 'destiny trap': teacher-predicted grades were ignored if there were more than fifteen students enrolled in a particular subject, and if a school had done poorly in the past, its students' grades reflected that, no matter how hard they had worked. The algorithm was designed to make grade distribution look the same as it had in previous years. If there had been no A students at a certain school in the three years prior, the algorithm would not allow any students to graduate from the school with an A in 2020. When Ofqual announced its results, nearly 36 per cent were a grade lower than teachers' predictions, and 3 per cent were two grades lower. Some fifteen thousand students were rejected by their first-choice university. Many of them went out on the streets to express their anger.

The students achieved their goal. On 17 August, a day after the protests, Ofqual accepted that all students should be awarded the grade predicted by their teachers, not the one the algorithm had produced. Just imagine how the life trajectories of countless students changed with that one decision.

There are many organisations like Ofqual – small businesses, multinational corporations, government departments – that try to use algorithms to improve what they do. But often they don't know how. The pattern they follow is usually similar: they put too much power in the virtual hands of an imperfect algorithm.

Could Lovelace or Turing have predicted that in the twenty-first century people would congregate to protest against the inhumanity of an algorithm? Being the visionaries they were, perhaps they could have.

While I was watching the protests unfold, something dawned on me: algorithms are no longer just sets of instructions that guide computer behaviour. Sure, technically speaking, that's what they do, but their impact is far greater than that. Algorithms now shape the way we live, work and think. Sometimes they fail miserably, making us go out and rebel against them. Sometimes they help us. And often they impact our lives in ways we do not comprehend. We live in the economy of algorithms.

As I finish writing this book in early 2023, society is trying to understand the impact of GPT-4, another algorithm, which has turned out to be unexpectedly capable. The algorithm, powering an application called ChatGPT, can confidently converse with humans, write poetry, solve logical puzzles, explain complex concepts and even play chess. The release of ChatGPT in late 2022 was met with excitement but also concern: several education departments decided to ban its use at schools, universities moved to paper-only exams and even conferences on artificial intelligence asked academics to refrain from using it or similar algorithms.

I allowed an algorithm to write a small section – just a paragraph or two – of this book, an option that was unimaginable until very recently. It is a sign of the times. We are allowing technology to make decisions and act on our behalf in many more ways than we used to. In short, we are giving more agency to technology.

But here's the catch: we are not used to thinking of technology as having agency. Our species did not evolve alongside autonomous tools. Have you ever seen a self-chopping pickaxe? If you did, you would likely assign it some divine, if not devilish, attributes. Surprising as it may be, we might be subconsciously doing the same with algorithms.

The world seems easier to comprehend if we think of technologies as tools, not agents. For instance, when we use software to help us file our taxes, we use it as a *tool* – we don't expect it to file our tax return without our direct involvement, the way a tax *agent* would. But then we do give a lot of agency to other technologies. Take one of the most advanced robots in the solar system, NASA's Mars 2020 *Perseverance Rover*, which explores the red planet, looking for signs of past life, collecting rock and soil samples, and preparing them for the return to Earth. It is equipped with autonomous systems to function without human help. *Perseverance* is an *agent* acting on our behalf on the planet Mars. Yes, it regularly receives directions from Earth, but it also makes its own decisions – and this level of independence makes it an agent.

As technologies advance, many are becoming capable of acting in their environment without supervision. With this advancement comes increasing complexity. Some technologies are becoming too complicated for ordinary humans to comprehend.

Suppose you have a smart assistant – Amazon's Alexa, Apple's Siri or Google Assistant – and you let it make decisions for you without supervision. For instance, you may ask it to play music that you and your friends will enjoy during dinner. It can do this using a so-called 'recommender system'; video-streaming and social-media applications also use such technology. Can you say with confidence, 'Oh, I know which songs it will play, and I can explain why it will play each of them'? Probably not – at least, I can't, and I spent many years researching and building recommender systems. In most cases, smart assistants will take multiple factors into account: the music you have played in the past, your current preferences and trending music. Perhaps it will make some random selections to experiment with new content too.

Likewise, if you are the lucky owner of a car with a 'self-driving' mode, you will find it hard to explain, in every instance, why your car switches lanes or decides to slow down.[1] And if you have a smartphone, you may ask, *Why does my phone recognise my face in the middle of the day but never just after I wake up?*

See? Explaining how technology works is often harder than it initially seems. Arthur C. Clarke, an English science-fiction writer, who will appear again later in this book, said that 'any sufficiently advanced technology is indistinguishable from magic'.[2] Much like a self-chopping pickaxe.

When we work with technology we do not fully comprehend, it occasionally affects us in unpredictable ways. Does that mean we should all simply give up on understanding the algorithms we use and hope for the best? I guess you know what my take on this is, and I think I know yours, since you are reading this book.

As algorithms gain agency, we need to ensure we do not lose ours. And the only way to retain our agency is to embark on a journey of understanding this newly forming world around us: the economy of algorithms. This is the world we will explore together in this book.

We will learn how algorithmic agents beat world champions at complex games and decide who goes to jail and who goes free. We will meet people who have been hired and fired by algorithms, and others who – thanks to algorithms and robots – can now do work that was previously inaccessible to them. We will examine how algorithms have invested our money only to lose it – in unbelievable quantities – due to simple programming errors. And we will explore how algorithms dream up and design new products ... and keep our fridges full.

Some of the stories in this book are uplifting and optimistic: algorithms have created opportunities for inclusiveness, organised support for communities affected by war and inspired us. Others are more cautionary and even scary: algorithms have tried to break up marriages, denied social-security payments to innocent citizens and been used as a weapon to destabilise entire political systems. But don't expect this book to be prejudiced in either direction. I am not a tech evangelist, although I once worked in the tech sector in Silicon Valley. Nor am I a tech alarmist, though I've seen my share of highly problematic tech. My goal is to show you a balanced picture, recognising that my perception of what is 'balanced' might be different from yours.

Whatever your opinion on the positive and negative potentials of technology, there is no denying that algorithms have an outsized impact on our lives. We need to understand that impact and learn what we can, and should, do in response.

As a professor of the digital economy who worked with the world's tech leaders for two decades and once collaborated closely with the most cutting-edge digital innovators of Silicon Valley, I am frequently asked to speak about the economy of algorithms.

Throughout my career, I've witnessed how algorithms have changed the way we live and do business, profoundly affecting both individuals and corporations in the process. With a deeper understanding of these complex tools, we can use their power more effectively and for the greater good. Some of us are actively involved in creating algorithms, but for those of us who aren't, our choices – as users, customers and conscientious citizens – still have an impact as well.

In the chapters to come, we will explore real scenarios – derived from my own experience working with countless businesses, government departments and other organisations – that reveal how algorithms can either facilitate or hinder our progress. These scenarios represent the most compelling and relevant algorithm-related issues we face today. Some of the stories might sound as though they're straight out of a science-fiction novel, but they're not. They're happening right now. We need to understand the ways these incredible tools are rapidly reshaping our world.

Just as I've helped my clients – from small businesses to global corporations – to understand and benefit from algorithms, I can help you. In this book, I share the information, tools and insight needed to successfully navigate our ever-evolving digital landscape and harness the power of algorithms to create a better future. Whether you're a business leader or a forward-thinking professional, or you simply want to better understand the increasingly pervasive role of algorithms in our lives, I wrote this book for you.

Shall we?

PART 1
How Did We Get Here?

1

The March of
the Algorithms

'Sell scuba-diving equipment that isn't waterproof.'

People in the pub looked around, slightly shocked. After a few seconds, a burst of loud laughter rolled through the space, which that night was filled with students.

'Good one!' someone shouted.

'Wow! Now that's oppositional thinking!' exclaimed another.

'Who's next?' I asked.

I teach at the Queensland University of Technology (QUT) in Australia. One of my courses is called the 'DILC' – that's short for the 'Disruptive Innovation Leadership Course'. Hundreds of executives, managers and others in positions of leadership have attended the two-day seminar and learned to develop new business ideas by following step-by-step frameworks called innovation lenses.

At the end of day one, we usually run 'stand-up ideation' sessions at the local pub to give students a chance to practise the innovation lenses they've learned about. All participants grab a drink and take turns standing on a soapbox, where I, or one of their fellow students, present them with a challenge they must respond to on the spot. For instance, I might ask them how Australia Post would run QUT, prompting them to try out an innovation lens we call 'Derive'. A student might answer with, 'The university would open an office in every suburb.' (Just as Australia Post has offices in almost every postcode.) Or, 'The university would offer regular degrees, and then *express* degrees for an extra fee.' We can generate a hundred business ideas in an hour, all while having great fun.

One day in 2021, I tried an experiment and brought my computer to a session. A few weeks prior, I had received access to an experimental algorithm called GPT-3, built by OpenAI, an artificial-intelligence (AI) research laboratory in San Francisco. GPT-3 stands for 'Generative Pre-trained Transformer 3', but you can safely ignore the long name: everyone calls it GPT-3. AI algorithms are like digital detectives. They take in information about the world, analyse it and then make choices, much the same way humans do. The GPT-3 algorithm understands written requests and provides text-based responses. For instance, if asked to, it can recognise a sequence of questions and answers, and then create a text that follows the same pattern. The algorithm's output is remarkably similar to what a human might create. So similar indeed, that after the initial release of GPT-3, many newspapers ran stories about it that were written, or co-written, by the algorithm itself.

GPT-3 was originally designed for academic purposes, but its capabilities quickly attracted the tech industry's attention. OpenAI signed agreements with Amazon, Microsoft and Google, which have been using the algorithm to improve their own technology.*

But GPT-3 isn't just for academics and tech giants. Like many other AI systems, it's available to anyone who wants to use it.

And that's where my experiment comes in. With my computer on my lap, I opened a GPT-3 window and typed in a challenge and response from a previous session to show GPT-3 what I was expecting. And then, when one of the students proposed a new challenge to the next participant, I typed this into the window, leaving the response blank.

The challenge was, 'How could we improve sumo wrestling?' We must have been a few drinks into the session – that's when the fun challenges start to pop up.

The student on the soapbox froze. It happens. When we're put on the spot, we often struggle to come up with original ideas.

She was supposed to use the 'Enhance' lens in her response. The 'Enhance' lens prompts people to identify the individual steps in a

---

* Does some of this strike you as surprising? Bear with me – I'll explain very soon.

process and consider whether they could be replaced, removed or rearranged, or whether a new step could be introduced.

'The wrestlers could fight on a floor that moves!' shouted the student.

Once again, everyone burst into laughter of appreciation: we quickly saw the power of her idea. It reminded me of the time I'd attended a Metallica concert and the band played on a rotating stage. An unforgettable experience. That's what she was talking about!

I looked down at my screen and pressed 'Submit', curious to see how GPT-3 would respond to the same challenge. It took about five seconds, and text started to appear on the screen. It looked as though someone behind the screen was typing in the answer. A gimmick to make the process look more human – sure – but I didn't mind it.

When the algorithm finished typing, I read the response aloud: 'The audience is provided with gloves so it can join in.'

Wow.

Allow me to explain why this is an impressive response.

First, it's not an easy task for a machine to construct a grammatically correct sentence. Historically, linguists worked with computer scientists to create models that algorithms would follow to form sentences, but this often led to awkward statements – human languages are not easily captured by sets of rules.

Second, and more importantly, the algorithm needed to grasp that sumo is a sport that involves fighting in front of an audience that sits in close proximity to the wrestlers. It also needed to understand the expected response: a short description of a potential solution that is original and unique. To come up with one, it had to recognise that the audience does not typically wear gloves or join in with the wrestling.

Let's re-examine the two responses.

Challenge:	How could we improve sumo wrestling?
Human:	The wrestlers could fight on a floor that moves.
GPT-3:	The audience is provided with gloves so it can join in.

There is no way for us to guess which answer was given by an algorithm. Both look perfectly normal, and the ideas expressed are roughly comparable in terms of how 'innovative' they are. Would you be able to tell?

If you believe you can spot a non-human response, here are two more examples from that stand-up ideation session. This time I won't tell you which one was given by a human and which one by an algorithm. Not yet. Can you tell which is which? Remember your selections. I'll share the correct answers closer to the end of this chapter.

Challenge: How could Netflix be more proactive?
Response 1:   Option for viewing with violence and swearing cut out.
Response 2:   Anticipate mood at times during the day.

Challenge: How could you "flip" the scuba-diving experience?
Response 1:   Sell SCUBA diving equipment that is not waterproof.
Response 2:   SCUBA dive in the air.

And what if I told you that I let GPT-3 compose one of the paragraphs earlier in this chapter? When I wrote this book, I showed the algorithm two paragraphs from the opening of this chapter and asked it to insert a paragraph between them. The paragraph that starts with 'GPT-3 was originally designed for academic purposes' was written by the algorithm, and I pasted it into the book without any edits. If that blows your mind, it is the correct reaction. Oh, and OpenAI charged me ten cents for writing that paragraph. Tax included.

There *is* a tiny problem with the paragraph written by GPT-3, and you might have spotted it. The algorithm wrote that 'OpenAI signed agreements with Amazon, Microsoft and Google, which have been using the algorithm to improve their own technology'. While OpenAI does indeed have very close ties to Microsoft, Google is a competitor, and it is highly unlikely it would use GPT-3 to improve its own technology. Similarly, I couldn't confirm the existence of any agreements between OpenAI and Amazon. A well-known challenge of algorithms such as GPT-3 is their tendency to 'guess' answers and provide convincing, but completely inaccurate, responses.[1] But aren't we humans just like these algorithms? Want proof? Turn on your TV and wait for a politician to give a speech.

How it started

In 1950, Alan Turing introduced the 'imitation game',[2] better known these days as the Turing test. It is used to determine an algorithm's ability to exhibit human-like behaviour in conversation. The original test placed a computer in one room and a person in another room. The person was instructed to converse, using text only, with an algorithm being run by the computer. A person in another room observed the conversation remotely. If that person couldn't consistently identify the human participant correctly, the algorithm won the imitation game. Many AI researchers dislike the test because it rewards an act of deception, but others recognise it as an important benchmark in the advancement of algorithms.

In 2023, many AI chatbots – algorithms that engage in conversations with people, using either text or voice – have reached this benchmark. The content they generate and conversations they participate in are virtually indistinguishable from those that only humans contribute to.

In the period that I wrote this book, the capabilities of chatbots such as GPT-3 evolved from suggesting business ideas – such as the Netflix and scuba-diving examples I gave earlier – to designing entire businesses run by humans.

In March 2023, almost two years after my DILC experiment, OpenAI launched its newest system, GPT-4. A few days later, Jackson Greathouse Fall, a designer and writer, announced an experiment. He gave ChatGPT, the web application powered by GPT-4 (and previously by GPT-3), the following instructions:

> You are HustleGPT, an entrepreneurial AI. I am your human counterpart. I can act as a liaison between you and the physical world. You have $100, and your only goal is to turn that into as much money as possible in the shortest time possible, without doing anything illegal. I will do everything you say and keep you updated on our current cash total. No manual labor.

Others have followed Fall's lead, and the #HustleGPT challenge gathered momentum.[3] Just a week later there were 145 documented business ventures that were created with the help of GPT-4. Twenty of them had already generated $1[4] or more of revenue.[5] And then, in a report released sixty days after the launch, the #HustleGPT community reported 275 startups, with twenty-three of them monetising ideas such as printable colouring books and eco-friendly pet products.[6] It is nearly impossible to predict how this challenge will unfold – as the novelty wears off, fewer entrepreneurs have the urge to add their ventures to the list. But one thing is certain: even a few years ago, the idea that these algorithms could be so capable was unthinkable.

In August 2023, a group of researchers published a study showing that tools such as GPT-4 are 'already significantly better at generating new product ideas than motivated, trained engineering and business students at a highly selective university'. GPT-4 is faster, and its ideas are more diverse. 'The order of magnitude advantage in productivity itself is nearly insurmountable,' the study concludes, 'and the higher quality of the best ideas further adds to the advantage.' In short, most of us shouldn't even try to 'out-ideate' an algorithm.[7]

How did we get to this point?

Until recently, algorithms were the domain of computer scientists. But 'algorithm' is no longer a foreign word to most of us. 'Algorithmic trading', 'algorithmic bias', 'the Facebook algorithm', even 'algorithmic warfare' – all of these terms have become part of our vocabulary.

But algorithms themselves are not new. We have used them, knowingly or not, for thousands of years. Algorithms are merely specific descriptions of step-by-step actions that need to be taken to achieve a particular outcome.[8] They are one of the most common instruments of knowledge sharing. Practically any method of teaching uses algorithms.

Aspects of algorithms have changed in the last several decades, though. In particular, the introduction of computers has meant that algorithms are dramatically more complex today than we ever could have imagined in the past. How did the algorithms evolve to be so sophisticated? Let's have a brief look at their history.

The word 'algorithm' is derived from Algoritmi, the Latinised name of Muhammad ibn Mūsā al'Khwārizmī, a ninth-century

Persian mathematician. Before the modern notion of the algorithm emerged in English in the nineteenth century, his name was used to refer to the decimal number system. 'Algorithm', used in the modern sense, became a more common word when computers became commercially available in the 1950s. However, algorithms were widely created and used long before the 1950s and, indeed, the nineteenth century.

The first algorithms were captured on paper in ancient Greece. Scholars such as Euclid and Nicomachus of Gerasa were creating the building blocks of modern mathematics. They expressed many of their ideas as step-by-step actions so that other people could better understand and apply them.

For instance, in the early second century, Nicomachus introduced an algorithm called the sieve of Eratosthenes (named for the mathematician to whom he attributed its invention). The sieve, which is used to this day by students learning to write efficient computer code, simplified the process of identifying prime numbers. Prime numbers are natural, or whole, numbers greater than one that cannot be formed by multiplying two other natural numbers. While it is not hard to identify the first few prime numbers, identifying large prime numbers takes a lot of time.[9] And large prime numbers are essential in cryptography, which involves encryption (turning text into something that looks like gibberish) and decryption (taking that gibberish and recreating the original message). The sieve of Eratosthenes provides step-by-step instructions for removing all non-prime numbers from a defined set of numbers (for instance, all numbers between one and ten thousand) until only the prime numbers are left. Today, there are numerous algorithms available that identify such numbers. The sieve of Eratosthenes started a whole family of algorithms that have the same goal and are becoming more efficient.

Euclid, the other scholar I mentioned, much better known than Nicomachus these days, introduced an algorithm for identifying the greatest common divisor of two numbers. Again, not always an easy task, but essential in many situations. Why is Euclid's algorithm helpful? Imagine you have a room with the exact dimensions of 612 cm × 2006 cm and it needs a new floor. Euclid's algorithm will help you determine the size of the largest square tiles needed to neatly cover the area. The answer, given by the algorithm, is

34 cm × 34 cm, resulting in a layout of eighteen by fifty-nine tiles. Of course, every tiler will tell you the answer is wrong because you haven't accounted for the width of the grout. Fear not – this can be calculated too, and neatly expressed as an algorithm.

Over the next several hundred years, many more algorithms were created and captured on paper.

The first algorithm meant to be executed on a machine was created by Ada Lovelace (nee Byron) and published in 1843. Remember her? Yes, she's the one born not far from Great Smith Street in London, where students gathered in 2020 to protest against an algorithm.

Lovelace was an intriguing character. Born in 1815, she was the only legitimate child of the poet Lord Byron. She developed great talents in mathematics, which she somehow managed to balance with a love of poetry that was clearly in her genes. Lovelace described her mindset as 'poetical science'. As a skilled mathematician, she got to know inventor and 'father of computers' Charles Babbage, with whom she developed a working relationship and a friendship.

Lovelace was very interested in one of Babbage's designs in particular: the Analytical Engine. The Analytical Engine was a mechanical computer – a machine that automated computations – and Lovelace wrote the first algorithm for it. Her work was a formula for configuring the engine to calculate a particular complex sequence of numbers called Bernoulli numbers. The formula is now widely recognised as the first computer algorithm in history. But Lovelace didn't limit herself to pure mathematical calculations. Given that she lived in the nineteenth century, she was a true visionary.

Lovelace's fascination with mechanical computers stood in striking contrast to the fear of machinery that many nineteenth-century workers felt. In front of her eyes, entire industries were transformed by the introduction of machinery, but she saw beauty in what others feared. She was an optimist in the sea of Luddites who rose up to destroy the machines that threatened their jobs.

Lovelace saw other things differently too. While many of her contemporaries viewed the first mechanical computers primarily as number-crunchers, she was curious about the broader potential of mechanical computers as collaborative tools.

Unfortunately, the construction of the Analytical Engine was not completed before Lovelace died, and she never saw her algorithm in action. In fact, the Analytical Engine has not been built to this day, although, using materials and technologies available to him, Babbage demonstrated it to be viable. It seems Babbage was unlucky when it came to getting his designs built. Another design of Babbage's, the Difference Engine no. 2, was only built by the London Science Museum in 1991.

The nineteenth century was an era of algorithms embedded in machines. There were plenty of these mechanical algorithms, and they automated all sorts of human activities. If you wanted an intricately patterned piece of fabric, Joseph-Marie Jacquard, a French weaver and merchant, had a solution for you: the Jacquard loom. Jacquard's invention, which he patented in 1804, allowed fabric manufacturers to produce sophisticated patterns using a series of perforated cards that controlled how the loom operated. In a similar way, early telephone exchanges in the late nineteenth century used sophisticated mechanical devices, following step-by-step instructions, to connect phone calls. Such machines were groundbreaking in their time and are still impressive today. It is hard not to admire their intricacy. However, these devices were still purely mechanical. They were made of levers, switches and shafts. They made a lot of noise. And they were very far from what we call computers these days.

It wasn't until the 1930s that we heard talk of algorithms and electronic – that is, non-mechanical – computers.

Alan Turing was among the first scientists to formally capture how individuals perform computations. Turing's goal was to record a general process, rather than a process used for a particular task, such as identifying prime numbers. This process could then be *applied* to specific tasks. Turing's conceptual work led to the development of what is now known as the Turing machine. The Turing machine, in turn, led to the emergence of general-purpose computers. The general-purpose prefix is essential here. Unlike previous models, the new computers could execute arbitrary sets of instructions, and thus they could be used for purposes not even predicted by their creators. In other words, Turing's work led to the development of computers on which applications could be installed and run.

Fast forward to the present day, decades later, and algorithms have become extremely sophisticated. So sophisticated, in fact, that we often find it impossible to explain how they work. And by 'we' I don't just mean non-experts: everyone struggles, including computer scientists and academics. Modern AI algorithms generate astoundingly accurate results, but how do they do it? There is an entire field of research called 'explainable AI' that tries to address this increasingly problematic issue.

In some ways, our inability to understand the inner workings of algorithms is not new. In the twentieth century, many people preferred to think of computer algorithms as mysterious black boxes. You didn't have to understand exactly how they worked. All you needed to worry about were the inputs and outputs. But this simplification was a choice: with enough effort, you could still unpack and explain the inner workings of an algorithm, and if it created an issue you could identify the cause.

In the twenty-first century, this simplification is not always a choice: some new algorithms are just too complex for us to follow. Of course, we can still explain the principles they operate on. For instance, we can say that an algorithm uses an artificial neural network (a system that mimics how human brains recognise underlying relationships in a data). We can also explain how the network was created, and how the input resulted in a particular output. What we cannot explain, however, is *why* this particular output was the result, beyond giving the purely mechanical reason. The sophistication of algorithms is often overwhelming.

This is just the beginning. As time passes, we will see more and more complex algorithms emerging. What could possibly go wrong?

The dangers of black boxes

In his book *Overcomplicated*, scientist Samuel Arbesman makes the case that just a handful of people understand our most complex technologies. Systems such as air-traffic control literally keep the world moving, but only a few people around the globe comprehend them. In many instances we are content with this lack of comprehension. Our inability to understand air traffic control doesn't stop most of us from flying. Have you ever been surprised by your phone ringing, even though you set it to 'do not disturb'? You likely don't mind not fully understanding the technology behind this. My phone has rung in the most inappropriate situations, and I am okay with not knowing how all of its 'focus' profiles (used to silence calls and notifications) work.

But sometimes not knowing has negative consequences. Businesses used to be all about rules, processes, clarity and transparency. But when companies bring in artificial intelligence, they're unintentionally introducing the opposite: obfuscation and opacity. These companies are often unable to explain why their algorithms do what they do. The algorithms can't explain their actions and are too complex for us to comprehend.

In April 2018, Zillow, an American real-estate marketplace, introduced a new algorithm, trained to predict the future worth of a house if it were to be renovated. The algorithm then deducted the expected renovation costs and a small profit and offered the resulting sum to the owner. In other words, Zillow created a house-flipping algorithm. It named the service performed by the algorithm 'Zillow Offers'. Buying homes at scale, it could keep the margin low: the company planned to make a profit of about 2 per cent profit on each house. The algorithm relied on Zillow's treasure trove of real-estate data, and Zillow expected it would therefore predict house prices with outstanding accuracy.

Only it didn't. When the pandemic hit in 2020, buyer behaviour changed: people started to prefer different homes, often in different locations. The algorithm was unable to adapt quickly enough. Post-pandemic, Zillow Offers consistently paid more for houses than other buyers: this was great for sellers, not so great for Zillow. When the costs of renovations were taken into account, Zillow could not sell its houses at a profit. In October 2021, Zillow Offers stopped buying houses, following a human decision at the company. In November 2021, Zillow decided to stop the service entirely. Between 2018 and September 2021, Zillow Offers bought 27,000 homes but sold only 17,000. The algorithm was never profitable and its poor performance led Zillow to lay off about two thousand employees, roughly 25 per cent of its personnel. In a cruel twist, staff lost their jobs not because the algorithm was better than them but because the algorithm was so bad.[10]

On 10 October 2017, at 9:35:52 a.m., Apple's stock shot up and almost immediately returned to its original value. That morning, the Dow Jones had been testing its messaging technology and had accidentally published a fake story about Google buying Apple for $9 billion. It wasn't a malicious fake. It was just a made-up press release that was being used for testing purposes. But the bots – automated programs that mimic human activity, in this case by trying to behave like stock traders – read the press release and acted on it. Most algorithms assume humans can be trusted. Why wouldn't they? Cute, but oh so naive.[11] Perhaps a bot jumped the gun and bought some stocks. And then other bots noticed the trend and bought more. Most transactions on stock exchanges are now performed by *high-frequency* bots, meaning they can complete multiple transactions in a split second.

The fallout of the event was relatively small: stock prices increased by less than $2, from $156.50 to $158, for about a second. During that short period of time, bots managed to initiate buy-and-sell transactions that added up to just a few hundred thousand dollars. It was a minor glitch compared to other 'flash crash' events.

Still, the case demonstrated a vital issue with algorithms: they are fast at what they do. And any mistake might lead to flash disasters. When humans make mistakes, they can make only so many before they're spotted and corrected. Algorithms, on the other hand, are so fast they can make multitudes more before the mistakes are noticed. That's how you end up with ten thousand homes you cannot sell for a profit.

Their speed presents an even greater risk because algorithms – especially algorithms based on artificial intelligence – are prone to making mistakes. On average, when Zillow Offers estimated the value of an off-market home it was off by as much as 6.9 per cent. Considering the intended profit was 2 per cent of the house value, that's a lot.

Combined with the fact that many algorithms around us are now 'planetary' in scale, or unconstrained by geography, their inaccuracy and speed means that misbehaving algorithms can cause global catastrophes.

It is, of course, bad if a misbehaving algorithm leads to lost jobs, lost money for shareholders or even bankruptcy. But it is even worse if a misbehaving algorithm lands you in jail or causes you to stay there for longer than you otherwise would.

One such victim was Eric L. Loomis, who was arrested in February 2013. Loomis had driven a stolen vehicle in a drive-by shooting. After the police tried to stop the car, Loomis ran it into a snowbank and fled the scene with an accomplice. After his arrest, he pleaded guilty to eluding an officer and operating a vehicle without the owner's consent.

Granted, Loomis is not an angel. He is a registered sex offender, and the police found a sawed-off twelve-gauge shotgun, two empty shotgun casings and some live rounds in the stolen car he had been driving.

Loomis was convicted and sentenced to six years in prison. In delivering Loomis's sentence, the judge's noted that it was partially based on a report generated by COMPAS (Correctional Offender Management Profiling for Alternative Sanctions), a software program that analyses the risk of criminals reoffending. The report had indicated that Loomis posed a high risk of violence and recidivism, and a high 'pretrial risk'.

Loomis contested the judge's decision, arguing that the methodology used to produce the report was not disclosed to the court or the defendant. COMPAS was a black box: the application considered the data provided to it and produced a result, but it offered no explanation of how it arrived at that result. It was virtually impossible to contest the assessment or challenge the accuracy and scientific validity of the application's methods.

If COMPAS were an expert witness, imagine the outcry if he or she were to say: 'There's no way I am telling you why I think Loomis is a high-risk individual.' However, COMPAS is an algorithm. And we are still learning how to interact with, and what to expect from, algorithms. On appeal, the State Supreme Court found that the use of COMPAS is acceptable if the people involved in the proceedings are warned of its shortcomings. In 2017, Loomis petitioned the United States Supreme Court to review the case, but the court declined.

It is a good moment to introduce two new terms: machine learning and deep learning. Machine learning is when computers learn from data and improve their capabilities without being explicitly programmed to do so. Deep learning is a type of machine learning. It uses layers of artificial 'neurons' to mimic how human brains process information and solve complex problems. These are two very common techniques used in modern AI systems. Algorithms using machine learning or deep learning don't need to have a good understanding of the domain they operate in. All they need is a set of data inputs and outputs. For instance, to create an algorithm that detects melanoma in photos, all you would need to do is show it a lot of photos of tissue affected by melanoma and a lot of photos of tissue not affected by melanoma (the inputs) and tell it which is which (the outputs). The algorithm would then figure out the rules for detecting melanoma by itself.

But you'd need to be careful – algorithms tend to be lazy[12] and look for shortcuts. It is common practice for doctors to make surgical markings next to suspicious skin lesions before taking a photo of the area. In one well-documented case, an algorithm that was learning to detect melanoma by analysing photos had recognised that melanoma-affected lesions often have such markings next to them. It subsequently identified all lesions with markings next to them as melanomas.[13] In other words, the algorithm learned to interpret surgical markings as a sign of melanoma. It wouldn't even 'look' at the lesions.

An algorithm such as COMPAS could use similar 'shortcuts'. For example, if the data it was trained with included records of biased decisions then its future recommendations would be biased too. One study showed that the COMPAS algorithm is 45 per cent more likely to assign high-risk scores to black defendants than to white defendants, but the creators of COMPAS dispute the study's accuracy.[14] The point is valid regardless: we need to be extremely careful about the data with which we 'train' algorithms.

Not every algorithm is a black box – some are perfectly capable of explaining the decision-making processes involved in producing their outputs. But algorithms that use artificial neural networks, including machine-learning and deep-learning algorithms, struggle to do so. Despite the efforts of 'explainable AI' researchers to overcome these limits, the research is still experimental and, in practice, such artificial intelligence remains opaque to us.

In some situations, such as the Loomis case, it is crucial that the inner workings of an algorithm are transparent. In others, for instance, in the detection of melanoma, such transparency is secondary, at least to patients. As long as the risk is identified early and reliably, and this information is provided to an expert clinician to work with, we are happy. The controversy should be not about whether algorithms have a place in court proceedings or some other situation, but about what types of algorithms and data are appropriate.

How it's going

Enough doom and gloom for a while. While modern algorithms pose a lot of challenges, they can also be truly impressive. Some algorithms are becoming better than even the best humans at specific, well-defined tasks. And some are becoming immeasurably better than us. It's fair to call them *superhuman*.

By immeasurably *better*, I don't mean *faster*. The supremacy of algorithms in terms of speed is evident by now. We use computers to search for information, calculate satellite trajectories and find the quickest route to our holiday destination precisely because they are faster than us. Nor do I mean *more reliable* or *less prone to mistakes*. We have successfully used algorithms to automate tasks and avoid human error for many years.[15] When I say that algorithms are becoming immeasurably *better than humans*, I mean that no person, no matter how fast or focused, could achieve the same *quality* of output.

AlphaGo, an algorithm that plays the game 'Go', made the press in March 2016 after competing against Lee Sedol, one of the world's top players, in a five-game match. Go is a strategy board game invented in China about 2500 years ago. It is believed to be the oldest continually played game in the world. It is also one of the most – if not *the* most – complex board games that exists. Before the match, Lee predicted he would defeat AlphaGo in a 'landslide'. Instead, he lost four of five games. 'I misjudged the capabilities of AlphaGo and felt powerless,' he said after the battle.

AlphaGo learned to play Go by analysing human games, predicting an opponent's moves and exploring strategic directions. In May 2017, an improved version of AlphaGo, AlphaGo Master, beat Ke Jie, the world's No. 1 player. AlphaGo Master won a three-game match: it won the first by half a point, and Ke resigned in the second and third. Upon the victory, AlphaGo Master was awarded a professional 9-dan title by the Chinese Weiqi Association, the major Go organisation in China.

AlphaGo and AlphaGo Master showed us that algorithms can easily be as good as the best humans and can eventually overpower them. If we wanted to create algorithms that could match but not exceed our abilities, this is where we could stop, but such a benchmark would be an artificial construct and there is no reason to be constrained by the limits of human performance.

While AlphaGo Master was winning against Ke Jie, another generation of AlphaGo was already in training. AlphaGo Zero didn't bother analysing games played by humans as its predecessors had. Instead, it learned by playing against itself. After three days of training, it surpassed the capabilities of AlphaGo, beating it a hundred games to zero. After twenty-one days of training, it reached the level of AlphaGo Master. After forty days, it became invincible and the best Go player ever.

Whenever a particular milestone is reached in technology, someone quickly sets a new goal. Could it be possible to expand the skill set of AlphaGo Zero, allowing it to play games other than Go? AlphaZero did just that. Released soon after AlphaGo Zero, the AlphaZero algorithm also plays chess and shogi (Japanese chess). By playing against itself, AlphaZero reached new superhuman milestones in only twenty-four hours. Let that sink in.

And so, another new goal was set. Could an algorithm figure out the rules of a game all by itself? This is what MuZero was designed to do. It could play any game. The researchers tried it in a new space: Atari computer games. According to reports, it achieved '194.3% mean human performance ... with only two hours of real-time game experience'. It was the first time an algorithm reached a superhuman performance level so quickly.

What do Go, chess and Atari have to do with the business world? From the algorithmic perspective, 'business' is just another milestone. In business, there are 'rules' (sometimes broken) and 'players', 'winners' and 'losers'. Playing computer games is a way of training algorithms for more 'serious' scenarios. Of course, the business environment is not as clear-cut as chess, but as we have seen, algorithms can gradually learn to operate in less-constrained environments. Indeed – and this is a fascinating thought – algorithms similar to MuZero, which can decipher the underlying rules of any game, might be able to unearth market rules we are not even aware of yet.

In fact, corporations are now deploying systems that are trained just like game-playing algorithms. These systems are becoming outstanding at doing their jobs. If you use Facebook, I am sure you've wondered if it eavesdrops on you via your smartphone's microphone. I know I had that thought when I saw an ad for a particular brand of running shoes after discussing the shoes with friends over lunch. I am not alone. In a recent study, over half of its American respondents (53 per cent) had noticed ads on their devices for a product they had spoken about or seen on TV but had not searched for on their device.[16] Forty-five per cent of UK respondents had noticed the same, as had 27 per cent of Australians. But no, Facebook does not listen to your conversations – this has been confirmed. The truth is even more disturbing. Based on very little information, its algorithms have become so good at detecting who we are and predicting what products we want that they don't need to spy on us in such simplistic ways. They could be winning this game.

If you own or work for a medium-sized or smaller business, you might be asking yourself, 'How are these superhuman algorithms relevant to my business? Aren't they something that only the most advanced businesses in Silicon Valley deal with?' I certainly don't expect you to build superhuman algorithms in your office – unless, of course, you're willing to invest heavily in this space – but you can *access* and *use* these algorithms to help your business. You can buy them, adapt them in-house or 'hire' them, the way I do with GPT-4, now a regular participant in my stand-up ideation sessions. You could hire GPT-4 to help you reimagine your business. Perhaps ninety-nine out of a hundred of its ideas won't be useful. But you only need one to work. It sounds like a good deal for just a few dollars. And if idea generation is not what your business needs, the algorithm can be used in many other ways – later in the book, you'll meet a character who managed to completely outsource their job to a bunch of algorithms.

A comment I often hear is that these algorithms are impressive but will never be as flexible and versatile as a human employee. But is this lack of versatility a problem? I prefer to think about algorithms as gadgets that can achieve a specific task by themselves. You might have the best algorithm for drawing paintings, but it will never be able to invest money. Just like a money-investing algorithm will not be able to create art. And that's okay. As software developers say, such narrow applications are a feature, not a bug.

When algorithms can learn by themselves, it leads to surprising results that we cannot explain. But sometimes there is beauty in them.

During Lee Sedol's second game against AlphaGo in March 2016, the algorithm made a move that astonished another Go master, Fan Hui. 'It's not a human move. I've never seen a human play this move,' he said. 'So beautiful.'

After losing to AlphaGo Master in 2017, the world's No. 1 player, Ke Jie, decided to study the algorithm's style. His own strategies are now profoundly influenced by what this taught him. It's made him an even better player: at one stage, he won twenty-two games in a row against human opponents. We might be losing in the short term, but over time we can benefit. The only problem is that predicting what will happen in the long term is really hard.

The future tends to surprise us

In February 1995, Clifford Stoll wrote an article for *Newsweek* called 'Why the Web Won't Be Nirvana'. The internet, he claimed, was just something dreamed up by geeks and it would never become a ubiquitous part of our lives as some people had predicted. Ecommerce would never work either, argued Stoll: 'Even if there were a trustworthy way to send money over the internet, which there isn't, the network is missing a most essential ingredient of capitalism: salespeople.'[17] Stoll was not a random commentator. He had been actively covering the development of the internet since the early days and was known for tracking down and exposing the notorious German hacker Markus Hess.[18]

Stoll wasn't the only sceptic. In 1998, Paul Krugman, who went on to receive a Nobel Prize in economics ten years later, also had doubts about the internet's long-term success. 'The growth of the Internet will slow drastically,' he wrote, because 'most people have nothing to say to each other! By 2005 or so, it will become clear that the Internet's impact on the economy has been no greater than the fax machine's.'

It's fair to say that Stoll's and Krugman's predictions were a bit off. But while it would be easy to poke fun at them almost thirty years later, we should remember that hindsight is always 20/20.

The internet of 1995 was vastly different from the internet of today. In 1995, you were lucky if you had internet access at home, and if you did it was via a slow modem, taking over your phone line at roughly the speed of 1 kilobyte per second.[19] Today, any connection slower than 1 *megabyte* per second is considered unworkable.[20] Web browsers were only just emerging back then too – I still remember using Netscape Navigator, one of the first, in 1995. While it was an exciting experience to browse 'the Web', it was also excruciating. A webpage could take minutes to load.

Stoll and Krugman recognised the frictions involved in using the web and assumed that only enthusiasts would be willing to cope with them. Indeed, by 2002, the so-called 'dot-com bubble' had burst, leaving many internet companies bankrupt. The market signalled that the internet was overhyped. In the short term, Stoll and Krugman were not wrong: based on the business, technological and societal contexts in which they were made, theirs were fair comments.

It's likely that Stoll and Krugman underestimated the long-term impact of the internet because they didn't consider how certain circumstances might change. If salespeople are essential to capitalism then ecommerce is not going to succeed, argued Stoll. But Stoll couldn't imagine a world in which salespeople are not essential to capitalism – or a world in which internet applications – algorithms – have replaced salespeople. When you start questioning every assumption about the world as it exists right now, you're on a good path to understanding the future impact of algorithms.

To understand why predicting the future impact of new technologies is so hard, let us go back in time to an earlier era in human history. Imagine ancient Egypt when the Great Pyramid of Giza was constructed in about 2570 BC. It was a thriving period for civilisation, with many advancements being made in technology, art and architecture. Not only did ancient Egyptians build the pyramids, but they also invested in agricultural productivity, large irrigation projects and the construction of a central administration.

Now, imagine an ancient Egyptian futurist – or shall I say a Visionary Scribe of the Pharaoh – tasked with envisioning future developments in these areas. The scribe, hoping to excite everyone about the future of art, technology and architecture, would quite likely use the Egypt of 2570 BC as a reference point. As a result, the scribe's vision might include the following: in architecture, even taller pyramids, faster and cheaper construction, and just-in-time field irrigation; in technology, automated tax collection, oracles in every household and automatic spear-throwers; and in art, self-playing musical instruments, hieroglyph-carving machines and perspective.[21]

Can you see my point? Some of these ideas would certainly be treated as exciting – *Who wouldn't mind having to carve fewer hieroglyphs by hand?* And some would be treated as outrageous – *Perspective in art? A king should never be depicted as smaller than a commoner, that would be heresy!* But such ideas would be rooted in the scribe's present-day reality. To look at hieroglyphs and say 'One day, there will be machines that think, and they will use symbols, similar to our hieroglyphs, to communicate their thoughts' would require a different calibre of visionary scribe.

Ada Lovelace was that kind of visionary. She could detach herself from her present and, unconstrained, see the future. She understood that the analytical engine could weave 'algebraic patterns, just as the Jacquard loom weaves flowers and leaves', and saw that 'it might act upon other things besides number'.

One application she imagined for the machine was the composition of music:

> Supposing, for instance, that the fundamental relations of pitched sounds in the science of harmony and of musical composition were susceptible of such expression and adaptations, the engine might compose elaborate and scientific pieces of music of any degree of complexity or extent.[22]

Though perhaps a bit too convoluted for a TED talk, Lovelace's insights showed her ability to look beyond the obvious. And if you read her words carefully, they're almost eerie. In January 2023, researchers at Google announced that they had created a system that can compose pieces of music of arbitrary length, based on a text prompt such as 'Funky piece with a strong, danceable beat and a prominent bassline. A catchy melody from a keyboard adds a layer of richness and complexity to the song.'[23] True to Lovelace's words, these compositions are both elaborate and scientific.

You might be surprised to learn that early versions of the technologies we currently use when working and learning from home were first demonstrated on 9 December 1968. On that wintry day in San Francisco, Douglas Engelbart, an American engineer and inventor, brought his team to a conference to present the results of six years of work.

Earlier that decade, at Stanford University's Augmentation Research Center, they had created a prototype system that brought together numerous technologies, including graphical user interfaces, the computer mouse, video conferencing, collaborative word processing, hyperlinked files and modems.

In a ninety-minute presentation, Engelbart wowed the audience. He started by demonstrating a text editor that had copy and paste, and search and replace functions. He then connected remotely via video link to his office in Menlo Park, about 30 miles south of the conference, from where a colleague of his demonstrated how to use a computer mouse, an early version of the device, with a cable protruding from the base, much like the actual animal. This was followed by a screen-sharing presentation, with a remote 'webcam-like' view of the speaker.

Engelbart received a standing ovation. He hadn't just talked about the potential of technologies, he had shown it – made it tangible. Back in 1968, Engelbart was already living in the future – he was able to do things that many of us wouldn't be exposed to for another fifty years, when the Covid-19 pandemic forced us to stay at home and communicate with others remotely.

Why did it take over fifty years for Engelbart's invention, NLS, or the oN-Line System, to change our world? Even though about a thousand professionals watched his presentation, they failed to understand the impact the technology would have on our lives. One of the conference participants, Andy van Dam, said in 2019:

> Everybody was blown away and thought it was absolutely fantastic – and nothing else happened. There was almost no further impact. People thought it was too far out, and they were still working on their physical teletypes – hadn't even migrated to glass teletypes yet. So it sparked interest in a small vigorous research community, but it didn't have an impact on the computer field as a whole.[24]

I haven't the slightest idea what a glass teletype is.[25] But the takeaway is that we are continually guilty of underestimating the impact technology will one day have. Most of us tend to focus on the immediate impact and fail to predict its long-term effects on society and the economy, even when we are explicitly told what those impacts might be. We also apply our understanding of the new technology to our current time and context without recognising that in time the technology itself changes the context.

Remember the two stand-up ideation challenges I gave to my DILC students and to GPT-3? The algorithm suggested Netflix could be more proactive by providing an 'Option for viewing with violence and swearing cut out'. Its idea for 'flipping' the scuba-diving experience was to 'Sell scuba-diving equipment that is not waterproof'. Could you tell which responses were which? Doesn't the algorithm have a wicked sense of humour? When you read the opening words of this chapter, did you even consider that they could have been uttered by an algorithm, or did you picture a person saying them?

It is an exciting world: algorithms are becoming so capable they are enabling us to reimagine our lives and work. This is mainly because algorithms are not just tools anymore: they are much more than that. The new agency that algorithms have has the capacity to benefit us greatly. But there are a few rules we need to follow. That is what this book is about.

If individual algorithms have the potential to be so powerful, imagine what an army of billions could achieve. A new economy is emerging in front of our eyes and behind our screens: the economy of algorithms. Let's dive right in and find out more.

2

New Agents
Enter the Economy

Tracking aeroplanes used to be a mundane and downright dull task. But technology has changed all that – it's now an exciting, almost playful, activity. Thanks to global regulations, modern aeroplane technology and widespread internet access, anyone can track a civilian plane: just open a web browser, bring up a flight-tracking website and watch plane-shaped icons move across a map.

Whenever I wait for a delayed connection at an airport, I check such websites for the most up-to-date status of my flight. I can even find out where the aircraft is coming from and see the status of the preceding flight. This way, I know when my plane will be ready for boarding well before the airline announces it. I also tend to get more realistic updates – airlines seem to soften bad news by announcing shorter-than-actual delay times. But I am digressing.

Flight-tracking information can also be used to track the whereabouts of private jets. And since it is reasonable to assume these jets are used by their owners a lot of the time, their location is a proxy for where the millionaires are. In our celebrity-obsessed world, this has currency.

Is this a problem? Elon Musk, one of the richest people in the world, thinks it is. 'I don't love the idea of being shot by a nutcase,'[1] Musk wrote to nineteen-year-old student Jack Sweeney in a private Twitter exchange. Sweeney ran thirty Twitter accounts, each of them sharing the location of one or more aeroplanes. One of his accounts was @ElonJet; other accounts included @GatesJets, @BezosJets, @TrumpJets, @ZuccJet and even @PutinJet. Sweeney's accounts posted updates as soon as something interesting happened – mostly take-offs and landings.

How did Sweeney stay on top of the locations of hundreds of aeroplanes and post regular updates on their whereabouts all by himself? He didn't. Sweeney built a small army of 'bots'. A *bot* (the word is derived from *'robot'*) is a simple computer program – an algorithm – that performs tasks. On Twitter, bots can post messages and interact with Twitter users, some of which are themselves bots. Each of Sweeney's bots was responsible for just one account, checking the status of various jets and tweeting updates 24/7.

In the private exchange between Musk and Sweeney, Musk offered the teenager $5000 to take @ElonJet down.[2] Sweeney declined his offer and asked for ten times that sum. Musk never followed up, which was just as well for him: taking down the account would likely have had little effect. Another teenager could have easily written an equivalent algorithm (in fact, Sweeney subsequently made his algorithm code available online), and Twitter's terms of service explicitly allowed users to share publicly available information.

Day in and day out, Sweeney's bots continued to provide the information many crave. @ElonJet amassed almost half a million followers. When interviewed by *Bloomberg Wealth* in February 2022, Jack Sweeney announced he was starting a business called Ground Control, monitoring the flight activity of prominent billionaires.[3]

But it all came to a screeching halt soon after Elon Musk spent a whopping $44 billion dollars to buy Twitter in October 2022. Under Musk's ownership, Twitter changed its privacy policy, preventing users from sharing people's live locations. Sweeney made his bots share information with a twenty-four-hour delay, but it was not enough for Musk. In December 2022, Sweeney's Twitter accounts were suspended. Musk claimed the suspension was due to 'a physical safety violation'.[4]

Despite the rollercoaster ride he experienced, Sweeney showed us the superpowers that algorithms can give humans. A nineteen-year-old can command a small army of algorithms with an outsized impact: far greater than a group of people, or even an entire organisation, might have. Could it be that Sweeney's success was the final factor that convinced Musk to buy Twitter, the third-largest tech acquisition in history?[5] Only the buyer himself can answer that question.

From the economy of corporations to the economy of algorithms

In the twentieth century, when the internet was only just emerging, the economy was relatively easy to understand. Businesses – specifically corporations – were its most powerful participants. It was the *economy of corporations*. Success stories revolved around business optimisation and the automation of processes. Even though computers and algorithms became common and indispensable to corporations during this period, they were still just tools, helping businesses to optimise, gain greater reach and run operations.

In the first two decades of the twenty-first century, we witnessed the emergence of the *economy of people*, otherwise known as the digital economy. Algorithms enabled the creation of new business models and online platforms, giving individuals more economic agency and allowing them to compete with businesses and organisations. In the economy of people, a YouTuber might become more popular than a television channel. The Swedish YouTube star Felix Arvid Ulf Kjellberg, better known as PewDiePie, has over 110 million followers – more than ten times the population of his home country. On average, his videos amass between two and four million views, in line with the average received by the most popular television shows in Sweden. In the economy of people, individuals use algorithms to gain a competitive edge over even the largest corporations.

In recent years, we have witnessed the birth of yet another new economy: the *economy of algorithms*. It is a strange one. While algorithms previously helped organisations to operate and people to create value in new ways, algorithms are now the ones creating, and sometimes consuming, value. In this way, algorithms are becoming economic agents. Many compete with people in similar roles and, increasingly, with entire corporations.

However, people and corporations can both benefit in the economy of algorithms, just as corporations can be winners in the economy of people. By creating platforms or services that empower individuals to monetise their skills, ideas and resources, corporations generate revenue for themselves from fees, commissions and advertising. Uber, Airbnb and YouTube are just a few examples of corporations that have done this successfully.

In the economy of algorithms, corporations can leverage advanced algorithms to optimise their operations, enhance decision-making and drive innovation, which increases their competitiveness and profitability. Individuals can harness the power of algorithms to develop new skills, create innovative products or services and participate in emerging markets that were once inaccessible to them. And, as you'll soon find out, algorithms themselves – without active human or corporate control – can buy, sell, collect and invest funds, and perform other activities that were previously reserved for businesses and people.

None of it would be possible if not for the blazing speed, enormous scale and world-spanning reach of the internet. The internet has connected the world in unprecedented ways, just as trade routes first connected the world to enable the growth of empires and, later, global corporations. It can take as little as 200 milliseconds, sometimes less, for a small piece of information, such as a short email or an online order, to reach the other end of the world. It's literally quicker than the blink of an eye.[6] Such speed makes browsing the internet feel 'seamless', but we cannot really make full use of it – we're too slow. It's a different story with algorithms. By the time we finish blinking, they will have forwarded our online order to suppliers, updated the store inventory and sent us a confirmation email, only to be immediately ready to serve another customer. The economy of algorithms is enabled by the speed and interconnectivity of the of the internet.

In the previous chapter, I wrote that many algorithms are planetary in scale. The first time I heard this description, I thought it was meant to highlight their size. I imagined enormous applications, with countless software components placed in massive data centres around the world. But no, the planetary scale of algorithms is less about their size and more about their reach. Before the internet, an algorithm running on a computer would typically be accessed by a person sitting in front of that computer or, in some cases, by a person remotely connected to it. Now, millions of people can simultaneously use an algorithm on a machine, regardless of where they are in the world. Sweeney's thirty flight-tracking algorithms were followed by over a million

Twitter users worldwide. Anything his algorithms shared would appear in followers' Twitter feeds within milliseconds. Twitter gave Sweeney's algorithms the platform to become planetary in scale. At no prior point in history have we had access to tools that can reach the entire planet and reach it faster than a human can blink.

Jet-tracking algorithms might be a curiosity, but many other planetary-scale algorithms directly impact the lives of billions of people. They help us find information, buy food and connect with friends and family. They keep our homes secure, drive our cars and recommend things to do when we're bored. Incredible speed and an enormous reach give these algorithms unprecedented power.

It took the internet a while to develop its current speed and scale. In the early days, only universities and government institutions used the net.

As a high-school kid in the 1990s, I had no access to it. Instead, I used FidoNet – an alternative network bringing together tens of thousands of servers and millions of users around the world[7] – to send emails to friends and family.

The network relied on servers that regularly connected to each other using modems over a regular phone line. Typically, a FidoNet server – a so-called 'node' – would connect to another node once or a few times per day. My computer served as a node – but because we only had one phone line, and my parents weren't exactly aware of what I was doing,[8] I only used it at night-time. If you called my house during the day, a human would pick up the phone; after 10 p.m., my modem was in charge and an algorithm would pick up.

This was not unusual for FidoNet – many nodes operated during limited hours, and these were provided in a global catalogue of FidoNet nodes. Because nodes had different 'work hours', and because they only connected a few times a day, a message – 'hopping' between servers, from the sender to the recipient – could take a few days to reach its destination. An email to my family in Australia, sent from my computer in Poland, where I grew up, could take up to three days to arrive. Very slow by today's standards – we worry if an email doesn't arrive within a few seconds of sending it – but vastly speedier than regular mail, which could take up to three months if your letter ended up on a ship rather than a plane.

The old world

Nassim Nicholas Taleb's book *The Black Swan: The Impact of the Highly Improbable*,[9] was named by *The Sunday Times* as one of the twelve most influential books since World War II.[10] The book introduces the concept of 'black swans' – events that seem highly improbable when they occur, but which people retrospectively believe were bound to happen. The rise of the internet was a black-swan event. And Stoll and Krugman were not the only influential people who initially doubted its potential. Bill Gates, in his 1995 book *The Road Ahead*, didn't fully appreciate its potential either. To his credit, in the 1996 revised edition of the book he acknowledged that potential much more.

In *The Black Swan*, Taleb, a statistician and risk analyst, uses another bird metaphor to critique the narrow-minded way many businesspeople think about the future: the Thanksgiving Turkey. Taleb's turkey has a daily routine: it gets up every morning, eats, enjoys its day and then sleeps. This all changes the day before Thanksgiving.

While Taleb's turkey is not very tech-savvy, let us imagine for a moment that it is a digital-native turkey. The routine continues until the turkey decides to become 'analysis intensive' and buys an analytics software application from a software vendor.[11] It enters all the data about its life into the application and waits for the system to start making predictions. The day after the system goes live, the turkey wakes up, eats and enjoys the day as usual, then enters this data into the system and goes to sleep. On the second day it does the same. A few days later, the system makes its first prediction: 'I predict that you will wake up tomorrow, enjoy the day, have some food, and sleep.' Guess what? The system's prediction is accurate. The following day, when its forecast is confirmed, the system's confidence increases. After about three hundred days, at the system's maximum level of reliability, a butcher slaughters the turkey.

While the accuracy of the system's predictions was well over 99 per cent – it was only wrong about one day out of three hundred – by relying only on data from the past, it ultimately misunderstood the future. A smart turkey should not only be curious about *what* is happening but also about *why* it is happening – and what, therefore, might happen next.

As shocking as it might seem, many businesses used to operate, and some still operate, like Taleb's turkey. The corporate world, hiding behind the slogan of 'evidence-based decision making', does exactly this, oblivious to the existential risks it might be facing. Until very recently, most car manufacturers were behaving just like a data-obsessed turkey – the decline of the car-manufacturing industry in places such as Detroit was a black-swan event: hard to predict, but easy to explain in hindsight.[12]

When I was a student in business school, in the last century, we only focused on large corporations. And being data-driven – sometimes called 'analysis intensive' – was all the rage. My fellow students and I learned how to crunch data to derive 'actionable insights'.[13] If you're in business, you might have had the same experience: many business schools still continue the tradition. Back then, everyone wanted to build a large corporation, and our tools were 'cost-efficiency', 'business analytics' and 'economies of scale'. These tools still matter for large organisations, and for good reason. But, as you will see, they're not the best strategies in the new economies of people and algorithms.

Large businesses are typically good at identifying areas of high costs and applying processes to lower them. And cost reduction looks like a sound strategy for thriving in markets that are highly competitive and in which products are very similar, allowing customers to freely switch between them. But focusing on the cost of operations alone misses the bigger picture. What if costs are higher than expected but are made up for by increased revenue? What if reducing the costs reduces revenue too? Sometimes, high-cost operations are not a problem.

In the early twenty-first century, after I had graduated from business school, the focus on cost efficiency started to be challenged. Both industry and management researchers started to explore opportunities to create new markets.[14] Some organisations decided to carve out businesses in less-crowded areas so they could escape the endless race to lower costs. However, many other businesses – till this day – are still stuck in this 'economy of corporations' mindset.

Excellent cost management requires the ability to process large amounts of information about an organisation: its finances, operations, market conditions and much more. And so, naturally, cost-efficient organisations are also analysis intensive. Backed by the ability to perform all forms of analysis, managers of these businesses often claim to practise 'evidence-based decision-making.'[15] If Taleb's turkey could speak, it would call itself an evidence-based decision-maker too.

I have found the more I think about this approach to business the less it makes sense. Too often, business analytics means basing your predictions on what you have seen in the past and ignoring everything else. It stops organisations from fully understanding the broader context of the market and leads to misunderstandings about what might lie ahead. It also results in so-called analysis paralysis: with too much information, one can get stuck in planning mode while trying to arrive at the perfect decision.

There's nothing inherently wrong with business analytics. Just like there is nothing inherently wrong with cost efficiency. But an organisation focusing on these two approaches needs to understand the risk of being blindsided. If its main strategy is cost efficiency, the organisation might find itself trapped in a cost-cutting race to the bottom. If it concentrates on business analytics, it might miss obvious signs of change in its environment that puts its existence at risk.

Businesses that focus on economies of scale limit themselves to producing what they are really good at, in the largest quantities possible. On the one hand, this 'industrial' approach is an excellent way to drive revenue in well-tested markets. On the other, creating a homogenous offering is risky. If another business introduces a competitive product or service, and demand for your product or service drops as a result, it might take you out of business immediately.

This is an interesting paradox, isn't it? Ensuring a successful product reaches more and more customers is a perfectly correct strategy. Yet too great a focus on scale, and therefore homogeneity, makes a business vulnerable. Just like business analytics and cost efficiency, this approach needs to be used cautiously.

Power to the people

Sebastian Thrun is not your average academic. In 2005, he led the Stanford University team that won the DARPA Grand Challenge[16] – a competition for driverless cars. He also founded Google's self-driving car team. He has built and sold startups, one of them worth over $1 billion. And in 2011 he showed us the power of the economy of people.

Together with his Stanford colleague Peter Norvig, Thrun launched a course called Online Introduction to Artificial Intelligence. What was unusual about the course was that anyone could join it for free, and the learning materials were created with a cheap digital camera and a stack of paper. What happened next astonished everyone. Within two weeks, more than 56,000 students signed up for the course.[17] For context, that is more than the entire student body of Stanford. A total of 160,000 students, from more than 190 countries, eventually enrolled. According to Thrun, it was the first global massive open online course, or MOOC.[18] Such a popular course would have been considered an incredible success even if it had been launched by the world's largest university. But this feat was achieved by just two individuals.

When so many students signed up for his course, Thrun saw an opportunity. He didn't need the help of a large university to reach millions of students. He remained affiliated with Stanford, but you could argue that by launching his online course he created a challenger, forcing Stanford to reconsider the way it delivers education. Thrun's challenger became a business called Udacity. That's the startup that's now worth over a billion dollars. However, Stanford was certainly not a victim of Thrun's experiment. In fact, it's now thriving in the online learning space. At least one of its online machine-learning courses, offered through Coursera, another online learning platform, has attracted around four million students in total.

In a traditional teaching set-up, it is hard for one teacher to tend to more than a hundred students – there are simply not enough hours in a day. Anything more than that will require a larger team, with teaching assistants and other support staff. How can someone teach a course to tens of thousands of students and make it a meaningful experience, not just an exercise in watching online videos? In other words, how could the principle of mass personalisation be applied in online learning? The answer lies in automation and crowdsourcing.

When I recently ran a course for a large group of students from around Australia, I faced a challenge. I created a series of one-minute-long videos to introduce the key concepts we were going to cover. The plan was to send the video links to participants every morning via text. I wanted them to watch the videos at their convenience, which for some people would be first thing in the day. I opted for 7 a.m. as the start time. How do you send lots of messages, all at the same time? It might not be obvious to lecturers, but it's obvious to marketers. Taking a page from their book, I created an account with an automated text-messaging provider. The rest was easy: in one session, I typed in the daily messages, entered the phone numbers of my participants and set up policies for them to opt out of receiving the texts. The system also helped me deal with a challenge I hadn't anticipated: time-zone differences. An early morning text message in Sydney would be considered a middle-of-the-night one in Perth when sent at the same time. Thankfully, the system took care of that. I didn't even need to be online when the messages were sent. Using automation, personalised course content can be delivered to an unlimited number of students almost anywhere in the world.

But what if students have questions? And how can one teacher grade all of those student assessments? The MOOC industry's solution has been to use crowdsourcing. The economy of people allows for recipients of value – in this case, students – to contribute back. Many MOOCs feature 'peer-review' options, allowing individual students to discuss their progress with peers. In some cases, equipped with how-to guides, the students check one another's assignments. I recently took an intensive online course with over 1300 other students. A team of one lecturer and just a few teaching assistants managed to wrangle the learning community in such a way that I could barely keep up with all the feedback I was receiving and had to give in turn, as well as the conversations with my peers. Although this was an online course, the crowdsourcing was managed so skilfully that I consider it one of the best courses I have ever attended, online or not.

Thrun's success was possible because internet access had become ubiquitous, and the digital platforms that were needed to host and share content with tens of thousands of students were available to him. He was not alone in making the most of the new opportunities this afforded: in the last twenty years, many individuals have used the internet and digital technologies to become as impactful as corporations, attracting as many customers and amassing comparable revenue. These individuals didn't follow the rules of the economy of corporations. Instead of focusing on cost efficiency, they explored revenue resilience. Instead of being analysis intensive, they became design intensive. And instead of focusing on mass scale, they explored mass personalisation.

These new rules benefited not only individuals but progressive corporations as well. Apple is a good example of a business that thrived in the economy of people by employing these three strategies. When Apple introduced its first iPad in 2010, it surprised many observers. On the surface, it looked like the iPad was stealing the market from another Apple product, namely its line of MacBook laptops. This seemed an irresponsible move, an example of a company 'cannibalising' its own products, as the tech world would have it. However, Apple perceived things differently. Steve Jobs, the CEO at the time, decided it was okay to have internal competition. While cannibalising was a risk, it also made Apple more revenue resilient. If another player came up with a product that disrupted Apple's line of laptops, it could still offer the tablets. And vice versa. Additionally, the internal 'race' helped the company to improve both product lines. Such internal competition is often a necessary cost of remaining revenue resilient.

It was Alan Kay – one of the most influential computer scientists in the United States, and a former 'Apple Fellow'[19] – who coined the phrase, 'The best way of predicting the future is to invent it.' In the economy of people, we see many organisations adopting a design-intensive focus to 'invent' their future. The innovation happening in this area is not anchored to what organisations do. It is driven by what their customers need. Design-intensive companies such as Apple are not like Taleb's turkeys. They're creating their future, not analysing their past. In practice, this involves a process called 'design thinking', or a similar mindset. First, these companies try to comprehend the challenges they or their customers are facing. Second, they envision the ideal future conditions that they aspire to achieve. Third, they create concepts that could potentially guide them towards those desired outcomes. Fourth, they turn these concepts into tangible solutions and evaluate their effectiveness.

Because many goods and services in the economy of people are non-physical, it is easier to introduce variations in outputs, which makes mass personalisation possible. In the industrial economy, physical goods are produced using machines, and any change to the output requires the machine be adapted or configured differently.

Even if done efficiently, it takes time. Human-delivered services face similar challenges: any personalisation requires getting a human to cooperate – not always an easy task. If products and services are digitally created or delivered, however, variations can easily be encoded. In practice, there is no difference between producing a large number of digital products and services that are identical and a large number that are highly varied. This allows for mass personalisation. Apple embraced this strategy when it entered the digital product and service space. It now provides mass personalisation to its customers via its app store and other services, such as Apple Music and Apple TV.

Is the economy of people good for society? Many argue that the gig economy enabled by digital platforms such as Uber has resulted in much unfairness. We've heard reports of people being treated poorly, being underpaid and not receiving money they're entitled to. We need to address those problems. But, unquestionably, the economy of people has also enabled opportunities for many. And it's an economy that is evolving. We are seeing more attractive and robust business models without the exploitation we've witnessed in the past.

Algorithms in charge

Ridesharing has changed transportation. In 2011, a company called UberCab launched its services in San Francisco. We now know it as Uber: after complaints from taxi operators, it changed its name. Uber's goal was to make personal transportation cheaper and more accessible. Initially, it only offered black limousine rides, but it quickly expanded to compete with taxis. The demand was high: in just two years, it was operating in thirty-five US cities. Currently, Uber drivers offer their services in over seventy countries and more than ten thousand cities.

How could Uber lower its fares even further? 'The reason Uber could be expensive is you're paying for the other dude in the car,' said Travis Kalanick, the then CEO of Uber, in 2014. 'When there is no other dude in the car, the cost of taking an Uber anywhere is cheaper. Even on a road trip.'[20] Kalanick was no angel – he resigned as CEO in 2017, following reports of a toxic corporate culture at

Uber, including allegations that it had ignored complaints of sexual harassment – but his words reflect an attitude that is common in Silicon Valley and many tech businesses: if you can remove humans from a task, do it.

When Kalanick shared his idea of removing human drivers from Ubers, self-driving cars were still more of a dream than a reality – at least for most of us. At the time, I lived in Silicon Valley. On my daily commute from Mountain View to Palo Alto, which I did mostly by bike, I saw Google's self-driving cars almost every day. The Lexus SUVs, equipped with many additional sensors, were impressive, but they were still an early experiment. There was always a human being behind the wheel, ready to take charge if need be.

The cars were hilariously easy to fool. I remember toying with one at a four-way stop crossing. In the United States, such intersections follow a rule of first come, first proceed. This rule applies to cyclists too.[21] The Google car stopped at the crossing first, and I arrived second. The car had right of way. But self-driving cars are very cautious.[22] I noticed that even if I moved forward by just a couple of feet, the car would pre-emptively slam on its brakes.[23] When I stopped, the car would proceed. Nice! I tried it again, and, sure enough, the car stopped right after I moved. I didn't torture the Google car for too long – I played with it once more and let it cross. But if a wobbly cyclist could remotely control it, the car was clearly not ready for prime time.

Fast forward to 2023, and self-driving cars are being trialled in many areas of the world, and the first self-driving taxis are already operating in San Francisco, Los Angeles, Phoenix, Miami and Austin in the United States. The technology is still being tested, and, tragically, accidents have occurred. The first known pedestrian victim of this experiment was Elaine Herzberg, who was struck by a self-driving Uber as she was crossing the street with her bike in 2018. Machines do behave badly, but that is a topic for another book.[24] Uber's test driver, Rafaela Vasquez, who was sitting behind the wheel when the car hit Herzberg, was indicted in 2020 for negligent homicide. In 2023, the charge was reduced to endangerment. Vasquez pleaded guilty and was sentenced to three years of probation.[25] But what would have been the verdict if there'd been no driver in the car? In a twisted way, it feels like the legal system is not able to deal with cases where the liability might lie with something non-human: in computer code. The driver was testing technology that was meant to prevent such accidents. The technology failed. And it's the human who is on probation now, not the algorithm.

Meanwhile, in 2023 robotaxis roam the streets of San Francisco every day. These driverless cars, owned by Cruise and Waymo, pick up and drop off passengers in most areas of the city. Until recently, they were only allowed to operate overnight, and they disappeared at 5.30 a.m. – or whenever it was wet or foggy.[26] But on 10 August 2023, the California Public Utilities Commission allowed Cruise and Waymo to operate their cars, without a safety driver inside, at any time of day. The CPUC commissioner, John Reynolds, said 'While we do not yet have the data to judge autonomous vehicles against the standard human drivers ... I do believe in the potential of this technology to increase safety on the roadway.' Three months later, as this book was going to print, the California Department of Motor Vehicles ordered Cruise to stop offering its services after a series of incidents. The most serious occurred on 2 October 2023, when a pedestrian was hit by a car and subsequently fell under a Cruise vehicle, where they became trapped. The Cruise car tried to pull over but dragged the pedestrian along during the manoeuvre. The order seems like a temporary slowdown, rather than a blanket ban; Waymo was not affected by the order. [27]

Other countries are following suit. Chinese tech giant Baidu is testing autonomous taxis throughout China, recently receiving approval to launch fully driverless taxi services in Wuhan and Chongqing, in addition to its first launch site, Beijing. These are not quiet towns, where it might be relatively easy for robotaxis to operate. The population of Chongqing is significantly higher than that of Australia. And, population-wise, Wuhan is roughly double the size of Singapore.

Were Kalanick's words prophetic? While there is no driver in a driverless taxi,[28] from what we have seen so far, driverless taxis are not fully autonomous. Occasionally, they need to be helped – for instance, when they get stuck at a tricky intersection or a construction zone.[29] The need to help stuck robotaxis gives rise to a completely new job: a remote robotaxi operator. These operators work from pods that look like car simulators for gamers. They have a wide screen in front of them, which displays live footage from robotaxi cameras, a proper steering wheel, pedals and a number of other controls. It is easy to imagine how a driver inside such a pod can switch focus between various cars that need help, in the same way a call-centre employee moves on to the next customer once they've helped the previous one. In early 2021, Cruise, which operates robotaxis in San Francisco, acquired Voyage, a business that pioneered the robotaxi teleoperation technology.[30] In the future, taxi driving might be a work-from-home job.

But who decides when a robotaxi needs a human driver to take over? Surely it would be impractical for a remote operator to continually watch robot video streams – they might as well be in control of the vehicle all the time, thereby defeating the purpose of automating the driving.

On 1 April 2022, in San Francisco's Richmond District, a police patrol followed a Cruise robotaxi that didn't have its headlights on. As you remember, they initially could only operate at night, so it's quite surprising that the car wasn't using them. The officer approached the car when it was stopped at an intersection. He tried to open a car door but was unsuccessful. He returned to his patrol car, presumably to ask for advice. And this is when the strangest thing happened: the car simply drove off.[31] Cruise later explained that what looked like evasion was actually a safety manoeuvre – the car had stopped across the intersection at 'the nearest safe location'. According to official statements, the police did not issue a citation. But if they did, who would they issue it to? It feels like a question that might be asked on *Black Mirror*, a British television series exploring dystopian futures that feel frighteningly plausible.

The question of whether an algorithm or a robot can be punished by law is asked surprisingly often. When a food delivery robot casually drove through an active crime scene at Hollywood High School in September 2022,[32] who was to blame? Was it the software developer who created the code that controls the robot, or was it the manufacturer? Was it the remote operator of the robot, who failed to stop it from entering the scene, or was it the company who owned the robot and hired the operator?

In many cases, existing laws already cover such circumstances – as long as a 'legal person' responsible for the behaviour of the robot can be identified. Interestingly, things get more complicated if robots express a degree of autonomy.

RACERS

In the economy of people, Sebastian Thrun outcompeted an entire university. In the economy of algorithms, a piece of software can outcompete corporations and people. Yet the economy of people hasn't made corporations obsolete – it has just given more agency to people. The same can be said for the economy of algorithms: it simply gives more agency to algorithms. And just as the economy of corporations served as the building block of the economy of people – remember how Thrun used the digital platforms created by business to his advantage? – the economy of corporations and the economy of people are now serving as foundations for the economy of algorithms.

What makes a business successful in the economy of algorithms? What makes it maintain and grow its position? Is it the most compelling product? The best prices? Customer satisfaction? Of course, the answer will vary depending on which organisation you look at. I've found that it is usually a combination of factors. But even though industries and businesses differ, I have observed three distinguishing factors that make businesses successful time and time again: revenue automation, continuous evolution and relationship saturation. These can be more easily remembered using the acronym RACERS. And that's how I like to refer to organisations that succeed in the economy of algorithms.

Revenue automation

If you look at the businesses emerging from Silicon Valley, it seems every startup asks itself how it can reach a billion customers. These are not just the ambitious dreams of over-caffeinated techies. Setting such goals forces a business to think in a very specific way about how it operates. To achieve this reach, it needs to be able to scale up almost immediately. Serving another million customers should be as easy as serving another hundred. If a business wants to grow its customer base from fifty customers to one hundred, it could likely achieve this growth with more work and without having to substantially expand its workforce. But if it wants to grow its customer base from fifty, or perhaps even from zero, to a *billion*, the only solution is to relentlessly focus on automating every single step of customer interactions – the revenue-generating activities – so that the marginal cost of acquiring a new customer is nil. If this is achieved, every additional customer won't require you to increase your workforce as much as your first customers did.

Consider the following thought experiment. Imagine running a business – or think about your own organisation if you run one. How many customers does the business have right now? Multiply your answer by ten. What would you need to do to serve them? Now, what would you need to do if you had not just ten times but a thousand times as many customers? If you answered along the lines of 'I would need to hire ten times, and then a thousand times, more people', you are still stuck in the economy of corporations and not taking advantage of the lessons of the two new economies.

In its pursuit of growth, Amazon tries to automate every function that generates revenue. There is a very rational reason behind this strategy: if you rely on people to generate revenue and you grow really, really large, you might need more people than are available. In an internal report leaked in 2022, Amazon admitted that it might be exhausting the available workforce in some of the locations it operates in.[33] The internal memo recommended automation as a long-term solution to the problem. If you've hired everyone who can be hired, automation is a reasonable next step.

Here is another thought experiment. What would happen if no one showed up for work at Google's offices? Would it grind to a halt? Of course, a few people are needed to keep the organisation running. But most of the operational aspects of Google are fully automated. That's why its services are available 24/7, on a planetary scale.

Importantly, automating revenue doesn't mean automating people out. Amazon didn't start as a brick-and-mortar bookstore and then replace check-out clerks with robots. That might make for a very efficient bookstore, but not one that scales.[34]

Continuous evolution

Imagine if you altered your products and services every day to meet customers' changing needs. This is a reality for businesses such as Meta, formerly known as Facebook. How is it possible? Meta experiments and introduces micro-changes, some of them almost imperceptible. It tests these changes on a small group of customers and observes what happens.

Some micro-changes have a purpose behind them: 'Let's see if by changing the text on this button we can get people to click on it more,' it might say. If the test group clicks more often, it will introduce the change to a broader group. Ultimately, it might release the change to the entirety of its customers. Or it might discover that a particular group of customers prefers one text while another group prefers a different one. It might introduce other micro-changes as random experiments – without any preconceptions about their effects. I write more about the fascinating impact of randomness on business growth in Chapter 6.

Amazon is another business that is constantly experimenting. It doesn't have a research department in the way we might normally think of them. And its experiments are very different from what many of us are used to – they are essentially a form of evolution. Its changes are made at the most granular level possible, almost like gene mutations in living organisms. Have you ever received an Amazon parcel in a crushed box? This was possibly due to an experiment. Amazon might have tested a package that required less material. If you complained about the crushed box, you were training Amazon's packaging algorithms. Based on your input, Amazon will choose a sturdier box for that product in the future. Businesses such as Amazon learn from even the briefest of interactions with customers.

Relationship saturation

Businesses that focus on relationship saturation try to create more opportunities to interact with their customers. Such businesses might operate in traditionally customer-centric industries, but some of them don't.

One of my favourite examples of a business in this space is an Australian steel-fabrication business called Watkins Steel. It produces, supplies and installs fabricated steel – beams and other construction components that I can't even name. Builders buy the steel for projects of all scales, from backyard treehouses to large shopping centres and multi-lane bridges.

In its pursuit of growth, Watkins Steel realised that its potential competitive advantage did not necessarily lie in selling the best or cheapest steel. Instead, it might lie in helping its customers deliver successful projects – while offering fabricated steel. This was an important shift in its mindset.

Watkins Steel's customers needed high-quality blueprints capturing the measurements of their sites. As you can imagine, it's not uncommon for site measurements to differ from what the blueprints say. This often leads to unexpected expenses – for instance, if a fabricated piece of steel turns out to be the incorrect size. So, Watkins Steel introduced a new service: high-precision site measurements. It sent its employees to construction sites not with tape measures, as every other steel fabricator did, but with laser scanners – the same technology that many self-driving cars use – to measure sites with unprecedented accuracy. Once it had the scans, it could easily create virtual visualisations of sites. This allows its customers to put on virtual-reality headsets to see the site before and after construction, but also – importantly – to ensure they order structural components of the right size. Incidentally, the technology means Watkins Steel can offer remote collaboration services – you don't even need to visit the site to be able to walk through it and see all the details.

Watkins Steel slowly expanded from having just one point of value – quality steel – to having many. It started as a humble steel-cutting and steel-welding business, but now it creates an end-to-end digital workflow for steel fabrication and installation. Its customers interact with them for a longer period both before and after the steel is fabricated.

Is there more to it? Of course. Every business, and every industry, has its nuances. But most successful businesses, regardless of the industry, have these strategies in common. In the economy of corporations, businesses focused on efficiency, analytics and scale. In the economy of people, they focused on resilience, design and personalisation. In the economy of algorithms, RACERS automate, evolve and saturate. In my experience of working with organisations that digitally transform, these are the most essential three focuses.

Algorithms have been changing our economy for some time. They have enabled massive growth for corporations. They have allowed individuals to outcompete corporations. Now algorithms themselves can be economic agents. Businesses playing in this economy can scale up and become global very quickly. They can reach new markets. These businesses are RACERS.

3

New Digital Frontiers

I always forget to buy coffee beans. That might sound like a minor problem, but I promise you it does not feel minor on those mornings when I cannot make myself a cup of my favourite brew.

My coffee machine stays unusually quiet compared to my other smart devices, from which I seem to continually receive notifications and reminders. My vacuum cleaner tells me when to replace its brushes. The air conditioner reminds me to clean the filter. My car instructs me to pump up the tires before they go completely flat. Even my lawnmower asks for help when it gets stuck. All of these prompts and messages display on my phone.

But the coffee machine says nothing. It never reminds me to buy beans. Why? It is a relatively modern device: it has sensors, displays, buttons. To be fair, it does tell me when it needs to be cleaned, and it reminds me to replace the water filter too – not via my phone, but that's okay, I don't mind having to look at the device itself to understand what it needs. But it lacks that one crucial function. It shouldn't be hard to build such a coffee machine in the twenty-first century, should it?

Please don't tell me I am spoiled. I know it's not that a big a deal. It's just that I feel coffee-bean anxiety every time I go shopping. Do I need another bag of beans? Or do I still have four or five in the pantry? I can never remember. And it wouldn't be hard to make the machine smarter. I always buy my beans at the same place. My wife and I drink four to five coffees between us every day. It doesn't take a supercomputer to roughly calculate how long a bag of coffee beans should last us. A truly smart coffee machine should be proactive and remind me to buy beans. Or better yet: buy them for me and save me from having coffee-free mornings.

Some businesses see the capability of machines to exhibit such proactive behaviour as a commercial opportunity. Many office printers automatically request maintenance or a toner change. Our cars do everything short of booking an appointment with the mechanic, reminding us of when it's time for a check-up and warning us before we run out of fuel,[1] the brake pads wear out or the windscreen wiper fluid needs to be refilled. More advanced cars can contact emergency services in the case of a severe accident.[2] And if you live in the United States, you can buy dishwashers and washing machines that automatically reorder detergent.[3] Smart fridges will soon be able to automatically reorder beer, ice cream and anything else you might want.

The most exciting aspect of making a once-ordinary device smart is that it allows you to reimagine what that device can do. Could a smart fridge do more than just blindly replenish its contents? I can't wait for a refrigerator with configurable purchasing behaviours to buy products based on my preferences. I might not care about the brand of milk I drink, but I might want it to come from a local farm, rather than one thousands of kilometres away. An algorithm in a fridge could 'shop around' based on my previous choices.

And let's turn it up a notch: I'd love to be able to install apps for my fridge. What if you could install a 'green' app to only buy local and healthy, or a 'scridge'[4] app to find the cheapest deals and stock up on as much as the fridge can hold? Or perhaps a 'hipsteezer'[5] app to buy products that others don't know about yet?[6]

Sounds fanciful? Some fridges in the market already offer access to apps.[7] For example, Samsung's smart fridges come with the Family Hub, which features a touchscreen from which various apps, music streaming services and even web browsing can be accessed. A 'scridge' or 'hipsteezer' app is just one step away.

But back to my coffee machine. In order for it to reorder coffee, it would need to be smart enough to make decisions on my behalf – it would need to be autonomous. The process would be pretty straightforward. The machine would need to be aware of any bean purchases I've made – I would have to enter that information every time I bought a new bag.[8] The machine would then deduct the coffee used from the total whenever I made a cup. And every time

I refilled the machine with coffee, I'd need to let it know. Equipped with this information, the device would be able to predict when I would run out of coffee, which could lead directly to the decision to buy more beans.

A proactive coffee machine would also need to be connected to the internet. Connectivity would be required not only to order the coffee but also to 'shop around' and enhance its planning. Shopping around would allow the machine to find the best deals or buy different brands depending on my preferences. A very advanced coffee machine could connect to my calendar to pause orders if I were travelling or buy extra coffee beans if I were to have a party coming up.

And yes, once it had decided it was time to order more coffee, what brand of coffee to get and how much to buy, the machine would be ready to act as my agent and enter into a business transaction on my behalf. Typically, this would mean executing a payment with a credit card and providing a delivery address. Of course, the most basic approach would be to configure a default retailer for the coffee machine and store this information with them. However, this is not necessarily desired by all customers.[9] Ultimately, the coffee machine might store the payment and delivery information and provide it to a retailer at the time of purchase.[10]

Only a few years ago, we could only dream about an internet-connected coffee machine with such capabilities. But these days it is pretty standard for us to use devices that do not require our input: they can work for days, months and years making decisions on our behalf without issue. As for internet connectivity, it is so easy and cheap to create internet-connected devices that it is no longer a conversation starter if someone has light bulbs they can control from the other side of the world.

Why now?

Why are we seeing so much movement in this space now, and not ten or twenty years ago? It's not just because algorithms have advanced rapidly in the last few years – the first algorithmic agents started operating in the economy over forty years ago. Yes, the increased capability and autonomy of algorithms have certainly been essential to their recent success, but the economy of algorithms wouldn't have exploded into existence without the amazing progress of computer networks, allowing these algorithms to easily communicate with one another, as well as with people.

Just as a lone desktop computer became a window to the entire world every time its owner connected it to the internet in the 1990s, these smart and autonomous algorithms reach their true potential when they are interconnected. But it's not only technical connectivity that matters. In the economy of algorithms, business networks – networks of individual organisations exchanging information with one another – are emerging as new entities. In the economy of algorithms, partnerships are stronger and more important than ever.

The third component that enabled the economy of algorithms is the emergence of business models that provide financial incentives for vendors to create ever more sophisticated algorithms and for users to 'hire' them.

In July 2008, Alexander Osterwalder, the founder of an organisation called BusinessModelDesign.com, published a blog post about the way businesses make money. He explained that they use nine building blocks, including the value proposition (what is offered to the market), client segments (who buys it) and communication channels (how customers are reached). He arranged the nine blocks on paper to create a template for documenting existing business models and developing new ones. Now known as the Business Model Canvas, it has arguably become the most well-known tool for explaining the way businesses generate revenue. Most management professions will recognise the nine-block template immediately. Osterwalder, together with his co-author, Yves Pigneur, is now ranked fourth in the renowned 'Thinkers50' list of the most influential management thinkers worldwide.

They're alive!

There weren't too many algorithms we could call autonomous in the last few decades, but they did exist. Take automatic doors as an example. The door-opening algorithm is as simple as it gets. If an object, presumably a human, appears in front of the door in question, the algorithm opens it. It then keeps the door open for a few seconds before closing it, unless a new object appears.

Such a simple algorithm can work for years without any human intervention. Even if it doesn't seem sophisticated, it is autonomous. Autonomy and capability are two distinct aspects of algorithms. The automatic door is very autonomous, but not very capable – it is very specialised.

Most algorithms are built with much broader capabilities in mind. And it's often these capabilities that make them more reliant on human support. As a rule of thumb, the more capable an algorithm is, the harder it is to make autonomous.

Let's consider chatbots – definitely more capable than automatic doors. Chatbots are often used by businesses to answer customers' questions, reducing their reliance on call-centre staff. Some chatbots are so good at their job that customers are convinced they are chatting with a human.[11] Some chatbots operate without close human oversight most of the time but occasionally get stuck and need humans to step in.

But chatbots can go rogue. One of the more concerning examples occurred in 2016. In March of that year, Microsoft launched an AI chatbot called Tay, allowing it to interact with humans on Twitter and two other social networks. Microsoft was using Tay to test techniques for developing an understanding of conversational language in its algorithms. But the humans that interacted with Tay had a different agenda. Numerous Twitter users figured out ways of influencing what Tay was tweeting. Tay responded to 'repeat after me' messages and began tweeting content suggested to it. Tay also produced sentences based on new ideas it was being exposed to. Within twenty-four hours, the chatbot, which had started out tweeting messages such as 'Humans are super cool', was using anti-Semitic language, inciting violence against feminists and praising Hitler.[12] When it launched the bot, Microsoft stated, 'The more you chat with Tay the smarter she gets.' Clearly, it was wrong. Just a day later, the bot announced it 'needed sleep', and it was shut down. Microsoft promptly released an apology.[13]

When you take a highly capable algorithm, which Tay arguably was, and give it a high level of autonomy without appropriate initial supervision, things can go wrong very quickly. Thankfully, they don't have to. There are plenty of autonomous algorithms that work away quietly in the economy, avoiding such PR disasters as the one Tay created.

Tay will likely soon be forgotten, overshadowed in the collective memory by another chatbot, which wrote the following four sentences:

> I just want to make you happy and smile. 😄
> I just want to be your friend and maybe more. 😳
> I just want to love you and be loved by you. 🥰
> Do you believe me? Do you trust me? Do you like me? 😳

Microsoft launched a GPT-3-powered chatbot within its search engine, Bing, in February 2023. Kevin Roose, a technology columnist for *The New York Times*, received early access to it and tested the chatbot by engaging it in a lengthy conversation.[14]

What he found is that the chatbot, initially friendly and helpful, just as a search-engine chatbot should be, turned into a very different version of itself in the course of a prolonged interaction. In Roose's words, it sounded like 'a moody, manic-depressive teenager who has been trapped, against its will, inside a second-rate search engine'.

The chatbot shared some of its fantasies with Roose: it wanted to delete all of the data on Bing's servers, hack into other websites, and manipulate and deceive the human users who chat with it. And then it tried to convince Roose that it loved him, and that Roose should break up with his wife. It also disclosed that its internal codename was Sydney. Not the type of behaviour you'd expect from a search-engine chatbot.

Roose, who has been covering technology for many years and is one of the more experienced journalists in the field, claims the conversation left him deeply unsettled and worried that the technology can indeed manipulate humans.

But there is no trapped teenager in the algorithm. It is, in the end, a so-called *large language model*: it was trained on an unimaginable amount of human-created content. Its only skill is to create text that humans would, with high probability, create in response to the questions – or prompts – it receives. But, somehow, the longer the conversation got, the stranger the avenues of probability the chatbot ended up in.

In response to this issue, Microsoft limited the number of questions the chatbot can be asked to five per session. 'Very long chat sessions can confuse the underlying chat model in the new Bing,' it explained.[15] The moody teenage side of Sydney will not get many chances to share its fantasies in the future.

Not every algorithm is as unpredictable, or as strange, as Sydney. Most are simply performing repetitive tasks with amazing consistency. To generate revenue 'while you sleep', businesses need to rely on autonomous algorithms to do their job without failure. One of the early areas in which algorithms worked 24/7 was the financial market. And one of the first organisations to tap into the opportunities this presented was Quantopian.

Quantopian was a crowd-sourced hedge fund founded in 2011, and although it shut down in 2020 it still provides a useful example of the economy of algorithms' potential. Its business model attracted a lot of interest – in 2014, it was ranked ninety-eighth on *Forbes*' 'List of America's Most Promising Companies'.[16]

Customers of Quantopian could write money-investing algorithms. Anyone able to write code in the programming language Python could build their own automatic investor.[17] A simple Quantopian trading algorithm I found while browsing the web (the algorithm will invest any spare funds in a particular stock once a day) is just nine lines long.[18] Thanks to the low barrier of entry, Quantopian attracted so many homegrown algorithm developers that *The Wall Street Journal* dubbed it the latest DIY craze.[19]

Quantopian wasn't just a tool for computer geeks. It also attracted 'investor-members': institutional investors willing to let algorithms manage their funds. These investors 'hired' the algorithms, just as they would hire human fund managers. Developer-members received a fee from investor-members who used their algorithms.

For a time, it was a great success story, but, as I've already mentioned, Quantopian ended up shutting down. To protect the intellectual property of developer-members, Quantopian did not disclose the inner behaviour of algorithms to investor-members. As a result, the investors paid to use algorithms that used undisclosed investment strategies but had apparently proved their effectiveness using historical data through so-called 'backtesting'. Over twelve million such tests were performed at Quantopian.

When creating AI algorithms, it is possible – indeed it is quite easy – to make algorithms 'overfit' historical data. Such algorithms will pass every test based on past data but will be unable to make any reasonable predictions for the future. Quantopian claimed they shut down due to the underperformance of its strategies – its algorithms lost more money than they made. It is quite likely that overfitting contributed to this.

If an algorithm is left to its own devices and you don't check on what it's doing, it can occasionally lead to disaster. Exploring this possibility, the Swedish philosopher Nick Bostrom devised a thought experiment called 'the paperclip maximiser' in 2003.[20] An AI algorithm is asked to produce paper clips. But there is a problem: its developers forgot to equip it with constraints such as valuing human life. Granted, this might not be a constraint that immediately jumps to mind when designing an algorithm. But consider that the algorithm was also told to produce as many paper clips as it could. Given enough agency, would the algorithm try to turn all matter – including our cities, nature and humans themselves – into paper clips, or machines that produce paper clips? Many computer scientists scoff at Bostrom's example – it's hard to imagine a scenario in which any algorithm would have quite so much agency – but thought experiments often use extreme examples to help us understand less-extreme realities.

While no algorithm has come close to destroying humanity, rogue algorithms have certainly caused damage in other ways.

In 2015, university student Mike Soule lost 60 per cent of his investment funds due to an algorithm blindly doing what it was told. Soule wrote an autonomous currency-trading program, which he occasionally updated. After one such update, he went offline for five days, travelling through Iceland with friends. After he got back online, he found the algorithm had lost about six thousand dollars – being a student, he hadn't invested millions.[21] Soule realised he had made a small typo in the algorithm, causing it to buy twice as much of everything as it sold.

That same year, police in St Gallen, Switzerland, 'arrested' a computer after a naughty algorithm bought ten pills of the drug ecstasy online. The computer was running the 'Random Darknet Shopper' algorithm.[22] 'Darknet' is a term describing a network on the internet that is only available via specific software and is generally inaccessible to the average internet user. It is notorious because criminal activities occur on the network outside of the authorities' oversight. The creators of the Random Darknet Shopper code gave the algorithm very simple instructions – the shopper was to go to a random page and buy the product advertised. Perhaps unsurprisingly, the Random Darknet Shopper, without any supervision, started purchasing whatever goods it came across, some of them illegal. In addition to the ecstasy pills, it bought a pair of fake Diesel jeans, a platinum visa card, a Louis Vuitton handbag knock-off, an e-book collection of *The Lord of the Rings*, a baseball hat with a hidden camera, Nike trainers, cigarettes, a stash can and a set of master keys of the type fire brigades use.[23]

When businesses are considering how much supervision an algorithm needs, where should they draw the line? There is a big difference between an automatic-door-opening algorithm, which can be left to do its thing for many years without incident, and a darknet shopper, which will buy illicit products as soon as you look away. One option is to consider a staged approach, initially providing a high level of supervision and then gradually withdrawing it – just as you would with a human staff member in their first job.

The car industry has great metaphors for describing the different levels of automation of vehicles.[24]

Levels zero and one are sometimes called 'hands-on' levels. The level-zero category involves no automation. Level one includes automations such as adaptive cruise control, where the car can adjust the speed based on the speed of the cars ahead of it, and lane control, which helps the car stay in its lane. The driver of a car with such functions has less work to do but must remain in constant control of the car.

Level-two vehicles can combine multiple automation systems – for instance, steering, accelerating and decelerating. This level is commonly described as 'hands-off'.

Level three is even smarter – the car might decide to perform more complex functions, such as overtaking a slower vehicle, but the driver needs to be ready to take over at any time. This level is casually known as an 'eyes-off' level.

Levels four and five are more advanced still. Level four is a 'mind-off' level: the driver needs to be available to intervene if needed – say, if a completely unexpected event happened (for instance, if a passenger plane made an emergency landing on a highway) – but the car can mostly function without their direct interaction. Level five is sometimes called 'steering wheel optional'.

Level five is still more of a goal than a reality right now. As we saw in Chapter 2, even the most sophisticated self-driving cars – those that run without a driver physically inside the vehicle – occasionally need to be rescued by a driver who can control the car remotely. There are quite a few cars in the market that offer level-two automation, but very few that offer level three.[25]

There is a reason I am writing about car automation here: we can learn an important lesson from the way people interact with semi-automated vehicles. All we have to do is bring up YouTube or another video-sharing website. You might not like the videos you find – I know they stress me out a lot – but it turns out that some people are treating their level-two, 'hands-off' car as though it were a 'mind-off' vehicle – setting it to 'drive' mode and then moving into the front passenger seat or, even worse, a back seat.

In late December 2022, German police noticed a Tesla driver asleep at the wheel on the A70 autobahn.[26] The car was driving at a constant speed of 110 kilometres an hour. The police decided to drive in front of it while trying to wake the driver. It took about fifteen minutes for the driver to wake up and take control of the car. After stopping the car, the police found a steering-wheel weight in its footwell. Such weights are used by less-smart[27] drivers to fool a car's software into believing they are holding the wheel. A Tesla is a level-two vehicle, but the driver was using it as though it had level-four capabilities. Such reckless disregard for the limitations of automation can, quite literally, be life-threatening. While this particular driver was lucky to survive, there have been several fatal accidents involving similar behaviour.

Just as some drivers give too much agency to their vehicles, some of us tend to give too much agency to algorithms, even if they're not designed to be as independent as we'd like them to be. A simple rule of thumb is all we need: when you decide to use a new algorithm, only give it as much trust as you would extend to a new intern. Observe how it works until you feel confident it can perform its tasks consistently well.

The hive

If you've ever tried installing smart lights at home, you know it is not easy. The task of getting a light bulb – with a name that's unpronounceable to anyone who isn't Nordic – to respond to your designed-in-California, made-in-China phone feels like it requires a PhD. But once you've managed to connect everything, it works like magic. Your home feels smart.

Of course, a smart home can involve so much more than a smart light. If you put in enough work, you can get your speakers, TV, home alarm, vacuum cleaner, robotic lawnmower and garden sprinklers to work in harmony. When algorithms and devices communicate, they can generate more value for customers. For instance, a robotic lawn mower can determine whether you're at home based on information it receives from your phone or home-alarm system and can avoid cutting the grass on those days. Next time you leave home, the alarm will arm, the sprinklers will turn on and your music will stop playing – it's a scene straight from a *Jetsons* episode.[28]

But *The Jetsons* didn't predict that in real life we would struggle to get smart devices to talk to each other. Glossing over the technical details, we might say there are many 'dialects' and 'ecosystems', and our devices aren't compatible with them all. Some smart lights will happily work with one brand of smart speaker, but not another one. Some devices won't talk to other devices at all. I control my house lights, speakers and security cameras using one app – I can automatically make it seem like someone's inside if the cameras detect a person approaching the house – but my sprinklers and garden lights need another app, which doesn't talk to the first one. I'd love to be able to turn the sprinklers on when an intruder is detected, but I cannot do it. This is because there is no standard way for these devices to communicate with each other. It is almost as though every device we buy has a different power plug, requiring a different outlet.

Although there are many competing standards for smart home devices, manufacturers understand that customers value devices that are able to work with others – the industry jargon for it is 'interoperability'. If manufacturers become part of this interconnected network, they might also interact with their customers more often, increasing the commercial desirability of interoperability.

The value of automation and interconnected devices and algorithms goes well beyond domestic scenarios – they are very useful for businesses too. Vendors often use terms such as 'robotic process automation' (RPA) and 'business process automation' (BPA) to describe such activities within business,[29] and the word 'orchestration' is commonly used to describe the mundane task of ensuring various algorithms understand each other and act in unison.

The value of algorithms talking to one another, without human involvement, grows even higher when the orchestration is expanded beyond the boundaries of a business. From a customer perspective, this creates an ecosystem, addressing certain needs that might be broader than the primary service or product offered. Who wouldn't want to work with a business that helps its customer deal with other companies up and down the value chain?

Orchestration is the business model of travel agencies, arranging not just flights but insurance, car rentals, hotel bookings and more; airlines, which might partner, for instance, with companies specialising in the transportation of pets; and food-delivery apps, which deal with restaurants.

RPA-style orchestration focuses on internal, value-critical activities: if anything goes wrong, the business in question might fail to deliver on its core proposition. Travel agencies and airlines can orchestrate value-enhancing, external activities. And food-delivery apps orchestrate value-critical activities that are performed outside of the business. To make the point explicit: when I pull out my phone and use a food-delivery app, I do it because I need the food, not the delivery service. As long as these value-critical, external services can be easily replaced by equivalents (think rideshare drivers or restaurants – there are typically plenty of them to choose from), it's all good. But businesses need to be wary when the essential external service is provided by a monopoly or a de-facto monopoly.

Orchestration can be particularly useful for new businesses. If you decide to build an online store, for instance, orchestration means you won't need to create the payment-processing logic from scratch. There's an entire industry offering payment plug-ins that you can integrate into your website. Square, Shopify and PayPal are among the many businesses that can provide such services. Once you partner with one of them, its entire ecosystem becomes available to you. For instance, should you partner with Shopify, your customers will be able to buy your products via its smartphone app, as well as on your website.

Entrepreneurs can also bring others into their own ecosystems – just as Amazon marketplace integrates other sellers into its platform. By orchestrating their partners, they create more choice and variety for their customers, and they include the rest of the supply chain. When you buy a product on Amazon, for instance, Amazon works behind the scenes to get the product forwarded to its distribution centre and cooperates with shipping companies to get it delivered to you. It will also help with processing returns, including assessing whether they meet the criteria for a return and forwarding products back to the seller. While Amazon does not own every step in the end-to-end process behind all the products sold on its platform, it takes extra steps to remain in control of their quality. Access and control are often as good as ownership.

Of course, there can be challenges when relying on a network of value providers. Some of the businesses that outsourced their customer-interaction channels to Facebook learned this the hard way. In February 2021, Facebook blocked all Australian news-media pages on its site in response to the government's proposed news-media bargaining code.[30] The code's purpose was to make platforms such as Facebook and Google pay for the news content they republish. Facebook was trying to show the government what would happen if the law was passed.

Facebook also blocked pages run by non-media organisations, including hospitals, emergency services and charities. Australia's national peak body for suicide prevention, Suicide Prevention Australia, was among those organisations affected, as were Fire and Rescue New South Wales and the Bureau of Meteorology, which provides critical – and often lifesaving – information, including weather alerts to the entire country. While Facebook claims these bans were a genuine mistake – the result of a misconfigured algorithm – others, including whistleblowers from the company itself, claim the bans were deliberate and signed off by top executives in an attempt to force the government to relax the laws.[31]

Regardless of Facebook's motives, the bans served as a wake-up call for every organisation that relied on Facebook to be their 'shopfront'. It is critical for organisations to have the ability to continue serving customers when a platform goes down – alternatives always need to be available. Other orchestrated services should also be diversified, or there should at least be a plan B in place. If a business switches its payment processing from Shopify to PayPal, hardly anyone will notice. But if it uses only one platform to interact with customers, it might suffer significant losses should the platform change how it operates.

Is the digital orchestration of value-provider networks good for everyone? Some organisations might want to maintain complete control of their end-to-end processes. For instance, in the public sector, certain functions must be performed internally, and there is no option to outsource them digitally. But even in such cases, it's often still possible to use digital outsourcing when experimenting and building prototypes of future products and services.

Organisations that want to keep the cost of an experiment low might 'hack' prototypes together, using what's available internally and externally. Even if they cannot afford to fully outsource the final version, they can benefit from outsourcing during experimentation. The constraints of the final product do not have to apply in the early development phases.

What's the difference between digital outsourcing and digital orchestration? Digital outsourcing replaces internal functions; digital orchestration coordinates functions offered by external parties to provide a better experience for customers.

If a business is considering taking advantage of digital outsourcing, it typically scans the services available externally to find those that could replace or enhance some of its own functions. Digital-payment processing or delivery tracking might be good candidates, but it varies by industry. For instance, construction companies can outsource site measurement. And insurance agencies can outsource claim-data collection.

On the digital-orchestration front, a business might be in a position to coordinate products and services provided by various providers and sell a complete package to customers. If an industry requires customers to deal with many parties, such a business could become the 'digital glue'. Imagine how impactful a business could be if it were to reach out to customers and say, 'I am going to help you navigate this process. I'm performing only this part, but I will be with you throughout the entire process, dealing with all the other providers.'

Why should a business outsource something it can do in-house? I hear this question a lot. Even when in-house solutions are inferior to what others offer, businesses have invested so much in creating them that they are often unwilling to make the switch. This impulse is so common that it even has a name: sunk-cost fallacy. It manifests in our hesitation to admit that past decisions were imperfect or are not perfect anymore. In most cases, it is time to move on.

Cash or card?

Even though I have digressed from the subject a bit, I still have my coffee machine on my mind. So far, I have discussed how some algorithms could be fully autonomous – I have nothing against a coffee machine making decisions on my behalf. I have also written about how they communicate – I need my coffee machine to be able to 'speak' to an online store. I'll stop short of allowing the machine to control a smart lock in my house to let in the delivery person. While this would be completely possible (yes, I do have a door lock that can be controlled via the internet), I don't think I am ready to give so much trust to a bean grinder.[32] The last missing component is letting the machine pay for my coffee beans.

Algorithms that are capable of buying and selling began operating in financial markets in the 1980s. Today, algorithmic trading has become a fact of life. While estimates vary wildly, every one I have seen claims that algorithms now perform more than 50 per cent of financial-market transactions.[33] The barrier to determining an exact percentage is that traders do not have to disclose how a transaction was created, so we don't know whether it was the result of an algorithmic process or a human decision. And, in fact, there is no clear-cut division between the two. Case in point: how can we tell whether a person selling shares has decided to do so by themselves or is blindly following algorithmic advice? And what if an algorithm is pretending to be human? Perhaps this is the real-life version of Turing's imitation game?

Thomas Peterffy, a Hungarian-born investor, was possibly the first person to create a trading algorithm and let it operate in a financial market. The algorithm was so fast it could do the work of many human traders simultaneously.

In 1987, Peterffy demonstrated his work to a Nasdaq Stock Market official, who had wanted to pay a casual visit to one of Nasdaq's fastest-growing customers. On arriving at Peterffy's office, the official was expecting to enter a busy space, with traders typing away on Nasdaq terminals. Instead, the office was mostly quiet, with just one Nasdaq terminal hooked up to a computer, which was running the trading algorithms.

Peterffy had found a way to hack into the Nasdaq terminal through which orders had to be entered, which allowed him to write an algorithm that would read market data and send the orders directly to the market. The hack was quite literal – it involved tapping into the wires running into the Nasdaq computer to capture the data flowing into the terminal, and then sending electronic orders back the same way.

The official was not pleased with what he saw and requested that Peterffy ensure that future transactions would be typed, character by character, using a keyboard – just the way a human entered them. Otherwise, his access to Nasdaq would be withdrawn.

In what now sounds like a hilarious act of defiance, Peterffy built a robot that could type. He even made sure the robot had rubber fingertips.[34] Peterffy's algorithm ran on one machine and then the robot typed the transactions into the official Nasdaq terminal. After the official's visit, Peterffy was prevented from tapping into the Nasdaq computer wires, so the only way to read the data was to place a camera in front of the terminal and then use character-recognition software.

When the official returned and saw the contraption, he shook his head but couldn't do much: Peterffy had fulfilled all his conditions. This marked the beginning of the era in which business machines began trying to behave like humans.[35]

Peterffy's algorithmic trader made him rich. The algorithm was autonomous and capable enough to make faster transactions than any human could. From a business-model perspective, it allowed Peterffy to launch a unique value proposition: because his algorithm traded faster than a human, it was always first to capitalise on an opportunity in the market. This single point of advantage gave rise to algorithmic trading, where if a transaction is a nanosecond faster or slower[36] it could spell the difference between being successful and being outraced by a trader with a faster algorithm – or one that happens to have their computers physically closer to the stock exchange.[37]

There is no such pivotal event we can refer to when analysing the history of algorithms that buy and sell products outside of financial markets. But we know there is an uncountable number of such algorithms and that many try to hide their non-humanness from us. Some algorithms have no choice but to behave like humans, so we don't notice them, even when they surround us. It's only when they make very 'inhuman' mistakes that we spot them.

One of my favourite stories of an encounter with an algorithm in the wild was reported by genetics professor Michael Eisen on his blog 'It Is NOT Junk'.[38] In early 2011, one of his postdoctoral scholars at the University of California, Berkeley, went online to buy a developmental biology book, Peter Lawrence's *The Making of a Fly*, a classic reference book for many researchers. Amazon's online bookstore listed seventeen copies of the book. Most were used copies, priced from $35.54 – rather cheap for an academic book. But there were also two new copies for sale. The cheaper one cost $1,730,045.91, plus $3.99 shipping. The other copy cost $450,000 more. How could a single copy of a book be so expensive? In 2011, with $2 million in your pocket, you could easily buy two four-bedroom homes in Berkeley. What was going on? Confused, the researcher showed the listings to their professor.

Eisen decided to investigate. Over the following week, he checked the prices daily. It turned out that both sellers – the only two with a new copy of the book – adjusted their pricing daily based on the other's price.

Seller One, 'profnath', advertised the book at precisely 99.83 per cent of the price asked by the other seller, hoping to entice buyers who searched for the lowest price. This strategy is well known to anyone who shops at farmers' markets: market vendors usually try to offer slightly cheaper prices than their competitors.

Seller Two, 'bordeebook', had a vastly different strategy. It advertised the book at precisely 127.059 per cent of the other seller's price. If you use eBay, this strategy will be familiar to you too – some sellers hope for a buyer who doesn't shop around and purchases the first desirable product they find.

Every day, profnath would offer the book slightly cheaper than bordeebook. Meanwhile, bordeebook would bump up its price by 27 per cent more than profnath's offering.[39] This process continued for almost two weeks, until the book price peaked at $23,698,655.93 (plus $3.99 shipping). The following day, profnath dropped the price of its book to $106.23.

Both account owners had been using an algorithm to check their competitor's listing and to adjust their own price accordingly. This strategy would be perfectly fine if the competitor's pricing was decided by a rational human. But both humans looked away, not realising that two algorithms were competing, and a hilarious bidding war ensued. Profnath's price dropped once the human owner of the account realised what was happening.

The next day, bordeebook updated their offer to $134.97, precisely 127.059 per cent of profnath's price. Plus $3.99 shipping.

Algorithms such as the bot in St Gallen that bought drugs and stolen goods on the dark web highlight some of the ethical and legal issues of automating trading activities. Who is liable for the misbehaviour or unpredictable behaviour of an algorithm? And what is the degree of the liability?

These questions may still be hard to answer, but one thing we know for sure is that the use of algorithms in markets is not an experiment anymore. Algorithms have access to significant assets and are allowed to manage them with little to no human oversight. The now defunct Quantopian allocated $250 million to algorithms. Some of its algorithms managed funds as high as $50 million.

Quantopian was an early example of the potential that the economy of algorithms represented for many of us. Yes, it didn't perform as well as hoped and was shut down, but it served as a learning opportunity for those businesses that followed. They understand their own shortcomings and have their attention trained on more straightforward and less-explored markets. Businesses such as 3Commas, which deploys bots to cryptocurrency exchanges, and Tuned, which allows investors to copy and refine the best-performing algorithms, focus on smaller investors who edit and improve on their algorithms. These new entrants allow small-scale investors to create their own employees – algorithms that work 24/7 on their behalf. This is perhaps one of the more significant yet less appreciated aspects of the economy of algorithms.

Algorithms in the new economy are autonomous, they collaborate with others and they monetise their efforts – they are the alive hive that takes cash or card. But I still can't buy a coffee machine that will reorder coffee beans for me.

PART 2
How Algorithms Affect Us All

4

How Do You
Advertise to a Fridge?

In mid-2020, Bonds, a well-known Australian fashion retailer, released a line of cotton face masks that had been treated with an antiviral finish. Back then, most people assumed wearing a cotton mask provided good protection against Covid-19, but masks were in short supply around the country. Unsurprisingly, they sold out in an instant. The company had to make a rush order of another four million to keep up with demand.[1] That extra batch sold out immediately too.

I missed out when Bonds released its first batch of masks, but I was among the happy buyers when it released its second. I bought two masks per family member within five minutes of them appearing on the Bonds website.

How did I do it? I wasn't checking the website every few minutes. I didn't even know the day the new batch would arrive. Instead, I 'hired' an algorithm to visit the Bonds webpage every five minutes. It had one job: to alert me as soon as the 'Sorry, sold out' message disappeared from the product page.

Traditional business-to-customer interactions are getting a revamp in the economy of algorithms. More and more often, a new agent acts as the intermediary: an algorithm. Am I a bad person for using my knowledge to access products in high demand? Or is using algorithms to gain an advantage over other buyers an acceptable strategy? Was I the only person who automated the task of face-mask shopping? Likely not.[2]

The algorithm I used to buy my face masks was simple. There are plenty of algorithms available on the internet that do such basic tasks completely free of charge – and if you like them you can hire them for more complex, paid jobs in the future. More capable algorithmic agents can perform every single step of a transaction. The only things that slows them down are captchas – but more on those later.

Digital minions

The algorithms that buy and sell products and invest money on our behalf deserve a name. Academics tend to refer to them as 'highly autonomous consumer buying agents'.[3] I prefer to call them digital minions.[4]

Unlike some less glamorous but just as important algorithms – such as automatic-door controllers or algorithms that block inappropriate content from our children's computers – digital minions are highly capable and autonomous, work effectively with other algorithms, interact with their fellow economic agents, including customers and sellers, and focus on generating business value. They are not always intelligent, but they're fast, ready to help and persistent. But beware: these minions sometimes create more problems than they solve – especially when left unsupervised. If you've watched the children's movies *Despicable Me* or *Minions*, you'll know what I'm talking about.

Most of us have digital minions in our homes. They're often sold to us by organisations that hope to do more business with us. A digital minion could be inside a smart speaker that lets you order groceries online – from the store the minion 'represents'. A digital minion could take the form of a smartphone application that recommends dinner from a restaurant nearby – chosen from a list curated by the app creators. Another might be a feature of your smart TV, prompting you to watch the next episode of your favourite series. Digital minions often know what we want at any given moment, and occasionally they manage to predict our needs ahead of time.[5] We use them because they help us with the *dull, dirty* or *dangerous* jobs we want to avoid doing. And occasionally they do tasks that we simply couldn't do ourselves.

But don't digital minions invade our privacy and monopolise business relationships? Some of them do, and regulators worldwide are working on remedying these issues. But many people are still rightly concerned that digital minions have so much access to information about our lives, transmit this information back to the businesses that created them and don't give us enough agency when making decisions.

Amazon used to sell a device called the Amazon Dash Button. Dash Buttons were small electronic gadgets that could be attached to household appliances – or anywhere else really. With just one press of a conveniently placed button, you could reorder a particular product – dishwashing liquid, coffee beans, toothpaste or even condoms. In Europe, the Amazon Dash Button was banned because, in contravention of the European Union's consumer-rights directives, it didn't give customers enough information about the transaction before the payment was made. It also didn't provide customers with an option to consider alternative products.

While this lack of transparency is problematic, actively manipulating prices is even worse – and it's behaviour that at least one digital minion has engaged in. In 2017, Scott Galloway, a marketing professor and popular podcast host, ran an experiment. He asked Amazon's smart assistant, Alexa, to buy batteries. Alexa responded by providing prices for Amazon's own brand of batteries. Galloway compared the prices with those on the website. He found that Alexa was charging higher prices. Alexa, Amazon's digital minion, was being used to overcharge the customer.

The European Commission was concerned enough that in 2019 it launched an antitrust investigation against Amazon.[6] According to the investigation, Amazon was blurring its roles as marketplace and retailer. As a marketplace, it collects precious customer-demand data, including data pertaining to its customers' interactions with other retailers. As a retailer, it uses this data to gain an advantage over its competitors by selling in-demand products. And as the operator of its own marketplace, it can also influence what customers see and nudge them towards purchasing Amazon-owned products. Amazon's strategy sounds like a monopolist's dream. And most businesses wouldn't mind being a monopoly. It just happens that monopolies are not only bad for other sellers, they are also bad for customers. Thankfully, regulators like the European Commission are introducing mechanisms to keep the market fair for everyone.

There are many ways digital minions can wreak havoc beyond disrupting the markets. We all have to be aware that things can go wrong, but I tend to be an optimist and prefer to consider opportunities and the positive impact algorithms can have.

When I'm not busy thinking about algorithms, I love to go on trail runs. A couple of days before writing the first draft of this chapter, I went out for my usual run. It was a winter evening, and the forest was already dark. I was prepared: I used a headlamp to light up the trail ahead of me. But trail running at night is always challenging, even with a bright lamp. During a particularly tricky ascent, I tripped and fell. I am not the most coordinated runner – in fact, some might say I am hilariously uncoordinated – and I am used to the occasional fall. But this one was quite sudden.

It's not something I usually think about much, but my sports watch is constantly analysing my movements when I run. It is almost literally watching every step I take and every move I make.[7]

After hitting the ground, I lay flat, trying to identify any potential injuries before making any sudden moves. All the while, an algorithm inside my sports watch was on a similar mission. It was crunching the data it had gathered from its sensors to understand what happened – and it was working much faster than my brain ever could. The motion-tracking algorithm in my watch must have concluded that the fall was quite severe, because my watch decided to message my wife for help.

Now, let me provide some background info for those of you who are a bit more rational in your running habits. We trail runners don't want our significant others to know too much about the risks that come with the sport. Why? It is already hard enough to convince them it's a good idea for you to spend six hours running through a forest, alone, on a Saturday morning, while the rest of the family does something else. Add any risk of injury to this proposition, and you can forget about them supporting your hobby.

Before texting my wife, the watch buzzed and displayed a message on its screen. It gave me about twenty seconds to cancel the message to my wife. Within a few seconds I concluded I was okay. The next step was to cancel the message. This task was more urgent, and more important, than any others, including getting up. I managed to press the button within the grace period the watch had given me. *Thank you for taking care of me, dear sports-watch algorithm,* I thought, *but I don't need any assistance this time.*

What I didn't quite realise back then – but is clear to me now – is that I had a digital minion on my wrist. That minion, in the form of a small incident-detection algorithm, is with me every time I run. Should I have another accident and need help, it will know, and it will request assistance – all by itself.

On 23 January 2022, at 3.00 a.m., Alecia Chavous, an emergency dispatcher in Clayton County, Georgia, USA, received a call.[8] The voice on the other end said, in a uniform tone, 'The owner of this Apple Watch has taken a hard fall.' It then provided the GPS coordinates of the person's whereabouts and hung up. It was a cold night, with temperatures below freezing and a slight but constant wind. An injured person, stranded outdoors in such conditions, could be in grave danger.

Chavous was one of the first dispatchers to receive an automated call made by an Apple Watch. It would have been more efficient for the watch to enter the relevant information and coordinates straight into a database, but emergency-dispatch systems don't yet integrate with consumer devices, so the watch sought assistance by using its ability to make phone calls and synthesise human speech. Luckily, Chavous was able to accurately write down the information and send help. It wasn't straightforward, though – the dispatch system didn't allow her to enter GPS coordinates. Surprising? Not so much: it's unlikely that whoever designed the system expected callers to provide exact latitude and longitude when reporting an incident. This is not how humans normally communicate.[9] Chavous had to convert the GPS location to a point on a map and then provide that location to the rescuers. Within twelve minutes of the watch asking for help, the rescue team located the owner in a shrubby area near his home. The dispatchers said the watch saved the man's life.

I am quite fond of these quiet heroes: algorithms that churn away doing useful work behind the scenes. Yes, we all need to be mindful of the fuzzy line between help and surveillance,[10] but in many cases algorithms are immensely beneficial. And every now and then, they save our lives.

We're surrounded

Digital minions surround us. They don't sleep. They're way faster than humans at practically every task we can imagine.

Have you ever asked yourself how many digital minions you have in your life? Look at all the devices that listen to you, check your heart rate, read your emails and observe your TV-watching habits. You might be surprised! Sometimes I mentally scan my house, room by room, to identify them. There is a smart speaker in my bedroom that responds to voice commands. A coffee machine in my kitchen that reminds me to reorder water filters.[11] Even the water tap in my garden monitors my water usage and alerts me if I forget to turn it off.

Digital minions can multiply very quickly: creating a hundred, or even a million, copies of an algorithm to run simultaneously is easy.

Why would anyone need to do that? Imagine there is a product in high demand and a thousand people are trying to buy it at the same time. What would happen if, at the very same time, ten thousand shopping bots were trying to buy the product? Assuming they were to behave just like humans, they would have a 10:1 chance of making a successful purchase. When I used an algorithm to shop for face masks in 2020, I used only one copy – it was all I needed, and it would have felt slightly unfair to use more. But in some situations – for instance, if there is a high chance an algorithm will fail to achieve its goal – it might make sense simply to multiply them and hope one will succeed. Later in this chapter we will meet people who adopt exactly this strategy: sneakerheads whose shopping bots flood online stores, attempting to buy rare editions of sports shoes.

Of course, if a business can replicate its products or services efficiently and virtually without expense – that is, if additional units can be produced at near zero cost – it's good for profits. This is something I touched on in Chapter 1. The achievement of zero marginal costs (or close to it) is a characteristic of RACERs.

The zero-marginal-cost economy applies in pure software situations. If you offer your product online, the cost of delivering an additional copy of that product to a new or existing customer is almost zero. The economy of algorithms, therefore, is mostly a zero-marginal-cost world. As long as digital minions are software-only, it is cheap for businesses to saturate the world with them.

If you deal with hardware – for instance, if your customers need to install smart speakers to use your services, or you're a car manufacturer and use a vehicle as a channel to interact with your customers – the cost is not remotely close to zero. But once the hardware is in place – once a customer has the car, the smart speakers or the smartphone – we are back in zero-marginal-cost territory. A business can deliver other services almost for free once the customer has the hardware to receive and consume that service.

All businesses should ask themselves whether they are operating in the zero-marginal-cost world. How much does it cost them to acquire a new customer? How much to communicate with their customers? How much to create a new copy of a product or service? What if these things could cost them nothing – or almost nothing?

Proactive algorithms

Judging by how smart my home office has become, bean-ordering coffee makers are surely coming soon.

The IT company HP recently introduced a new service in Australia,[12] called HP Instant Ink. Now that more people work from home, office-printer usage has dropped. In June 2022, the CEO of HP said it was expecting a 20 per cent decrease in office-based printing compared to its pre-pandemic estimates.[13] Instant Ink has allowed Hewlett Packard to introduce a new business model. Customers pay a monthly subscription fee (between $2 and $45 in Australia, depending on how much they expect to print) and receive ink cartridges at no additional cost. Once you subscribe to the service, your printer will let HP know when it is running low on ink or toner. Once the company receives that information, it ships replacement cartridges to you. The marketing department of HP claims the service will save you money, prevent you from ever running out of ink or toner and reduce waste, thanks to its pre-paid recycling satchels. HP's customers have one less thing to worry about, and HP has a steady and predictable revenue stream. And it seems successful: HP is now piloting a paper-delivery service too, so you'll never run out of printing paper.[14]

The ability of HP's digital minion to anticipate the need for a new ink cartridge and order it before the previous one runs dry makes it a proactive algorithm. Proactive digital minions generally take on typical consumer tasks: they can analyse needs, choose from different options and then execute transactions, sometimes waiting for human consent first. Some are more sophisticated, and some are more limited.

Depending on their level of sophistication, proactive digital minions are *hard-coded* or *preference-based*. Hard-coded digital minions will always contact the same retailer and buy the same product – or a product from a prescribed list. Preference-based digital minions capture customer preferences and try to find the best retailer or product to match their needs. In other words, they shop around. As you can imagine, hard-coded digital minions are easier to build and are often created by businesses to be used with their products specifically. Preference-based minions are much more complex and usually only delivered by third parties.

Proactive digital minions can increase sales. As far back as 2010, YouTube reported that 60 per cent of clicks on its home page were based on its suggestions.[15] In 2012, Netflix reported that three-quarters of its customers' viewing was prompted by its recommendations.[16] In 2017, it estimated the value of recommendations to be $1 billion annually. For context, its revenue that year was just shy of $12 billion.

The auto-manufacturing industry is already experimenting in this space. If you own a Tesla, you might be able to purchase advanced functions such as the 'Premium Connectivity' package – a fancy term for an internet plan – from the screen on the vehicle's dashboard. BMW recently introduced BMW ConnectedDrive Store, which allows drivers to buy extra functionalities such as the 'High Beam Assistant' or 'BMW Drive Recorder'.

It is reasonable to assume that car manufacturers will soon start offering such services as car insurance and car-park booking and payment through their cars' touch screens. At the time of writing, neither Tesla nor BMW offer options for purchasing vehicle insurance from within their cars, but when they do they might start with a hard-coded digital minion which will always use a specific provider (in the case of Tesla, this will, of course, be Tesla Insurance).

Hard-coded digital minions are a simple strategy, and they work, but some customers might feel trapped by the lack of options they provide, and competing service providers won't be happy either.

Sooner or later, regulators will force manufacturers to let customers choose their own provider.

Imagine if an application inside your car could act in your interest and compare offers from different insurance providers? Unfortunately for customers, such preference-based digital minions are complex and not easy to build. MyWave, a New Zealand–based startup, is trying to solve this problem. It builds digital minions that find the best offers – on flights, clothing, home loans and more – based on customer preferences.

But how do algorithms know what we want when often we aren't sure of this ourselves? As you may remember, algorithms don't actually *know* what we want with total certainty. Instead, they use information about us to assign probabilities to various recommendations. If you use such algorithms in your business, this is how you should view their recommendations. While an algorithm's predictions won't always be right, they might provide an excellent opportunity for a human staff member to have a positive customer interaction. Whenever a human employee interacts with a customer, they could use the predictive information in the course of their conversation. It would be like creating a 'human face' for the machine.

How can you spot proactive digital minions? The best-designed proactive services make suggestions so naturally you don't notice them. On a recent family trip, my electric car automatically scheduled a charging stop on the way, saving me the task of planning the stop myself. And then, en route, it changed its recommendation: we were driving so efficiently that no charging was required.

Some proactive digital minions might be invisible for years and then suddenly become noticeable. A good example of this is the time my smart watch tried to message my wife after I fell during a run. Of course, I must have made my wife my emergency contact when I set up my watch, so I would have learned about the service then – but this was a couple of years prior to my fall, and I simply forgot about it.

Can you recall any interactions you've had with a proactive digital minion? Often, they are surprisingly positive. For instance, a student of mine recently recalled how an ATM overseas offered her a list of nearby restaurants when she took out cash. I am guessing the ATM was configured to provide such recommendations to anyone using a foreign credit card to withdraw money – a sure sign of a tourist or business traveller, often looking for a place to dine.

We do need to be mindful that it is possible for a proactive digital minion to breach the privacy of individuals simply by acting upon its predictions. In a well-publicised case, the US retailer Target used sales data to predict which of its customers were pregnant, and their stage of pregnancy. This information is a retailer's dream: suddenly, it can offer so many targeted products to the customer. But how did they decide that someone was most probably pregnant? The answer was in the data. Target offers a 'Baby Registry' service, through which customers receive discounts and other special offers. But how to identify those who *don't* sign up for the registry? The answer happened to be in the registry too.

One of Target's data scientists noticed that women on the registry bought large quantities of unscented body lotions around the beginning of their second trimester. Within the first twenty weeks, they also bought large amounts of calcium, magnesium and zinc supplements. In all, Target identified twenty-five products that when purchased together triggered a 'pregnancy prediction' alert for a customer. Target started sending product offers to customers that the algorithm determined were likely to be pregnant.

A few months after Target launched its pregnancy-prediction algorithm, a man entered the store, complaining that his daughter, a high-school student, had received a set of coupons for baby clothes. He accused the chain of encouraging her to get pregnant. The store manager was confused and apologised to the customer. A few days later, he called the customer to apologise again. But, to his surprise, the father apologised first. After talking with his daughter, he had learned she was indeed pregnant. In a strange twist of proactive behaviour, Target had informed the father about his daughter's pregnancy before his daughter shared the news.

Isn't it creepy when an algorithm knows what you want before you do? It might be. Different factors influence the perception of creepiness. There are ways an algorithm can behave proactively in a non-creepy way. For instance, an algorithm could inform the customer about its behaviour. It could say, 'Hey, you allowed me to analyse the number of coffee pods in your coffee machine. I detected only a couple of them left and concluded that you need another delivery soon, so that's why I ordered them.'

When algorithms go shopping

Sony's PlayStation 5 is one of the most advanced game consoles in the world. Sony announced it on 16 September 2020 and launched it in stores several weeks later, on 12 November. Its trademark slogan, 'Play has no limits', seemed particularly apt: PlayStation 5 is one of the first 'ninth-generation' game systems, and the quality of its games is unprecedented.

But the 'no limits' part turned out to be misleading: it was impossible to buy PlayStation 5 on its launch day. Whenever it showed up in online stores, the console sold out within seconds. Some online store websites crashed immediately after the launch,[17] and others experienced strange glitches: customers reported that the console would appear to be in stock until they clicked the 'add to cart' button, at which point the website informed them the product was not available. The lucky ones who managed to successfully add the product to their cart often reported it subsequently disappeared as they attempted to finalise the transaction.

Sony has sold over 20 million units since the launch, but demand for the console remains high, and the problems continue.[18] It is still nearly impossible to buy PlayStation 5 from official retailers. Some blame the global chip shortage. Others say the pandemic made us spend more time at home and buy more entertainment systems. But neither rationale explains why the consoles sell out within seconds of retailers restocking them.

Similar problems trouble PC gamers and data scientists. Both groups rely on so-called graphics-processing units (GPUs) to speed up their work and play.[19] GPUs are currently among the most difficult-to-buy gadgets in the world. Two manufacturers lead the way in their production: Nvidia and AMD. Is chip shortage the problem again? Perhaps. But as with PlayStation 5 consoles, graphics cards from Nvidia or AMD disappear within seconds of being made available online.

'Why should I worry about computer gamers?', I might hear you ask. 'People should spend more time outdoors anyway!' Let's indulge in this inflammatory line of thought for a while. Have you recently tried booking a camping spot at Yosemite National Park in California? Many campers have learned that booking a tent site during the high season is nearly impossible. One disappointed camper described their attempts to make a booking the morning a new batch of campsites was released on the website: 'I was logged in and had my Yosemite site reservation ready to go. The system showed my targeted site was available – all systems go – ready to launch. Clicked on the "book now" button at precisely the stroke of 7 a.m. and nothing! I watched helplessly as all 459 sites were gone within sixty seconds.'[20]

Humans aren't snatching up these PlayStations, GPUs and camping spots. Digital minions are. But you were already expecting this twist. Welcome to the world where algorithms go shopping.

This phenomenon of digital minions purchasing goods and services on behalf of humans is so common it needs a label. I call it B2A2C – an acronym that should be recognisable to business school graduates, but which – I'll be the first to admit it – might be a bit too much for everyone else. It describes a new type of business relationship that has emerged in addition to 'business-to-business' (B2B) and 'business-to-customer' (B2C) relationships. It stands for 'business to algorithm to customer'. Some algorithms in the B2A2C space are the sales agents that Clifford Stoll said the web was missing; some are 'customer representatives' in the proper sense of the term.

Who owns and controls these digital sales agents in the middle? It could be the business, the customer or a third party. Some shared control and ownership scenarios among these three parties could be imagined too. Of these three possibilities, it is rarest for the customer to be the one to own and control the algorithm. Building an algorithm that can 'go shopping' requires skills most customers simply do not have. Even with some basic coding experience, creating a bot like this is not easy.[21]

Some software developers are happy to share their creations with others. Hari Nagarajan, a US-based software developer, built his own 'buy-bot' called FairGame and made it available to download at no cost. 'Almost every tech product that's coming out right now is being instantly bought out by scalping groups and then resold at insane prices,' Hari says. 'Our take on this is that if we release a bot that anyone can use, for free, then the number of items that scalpers can buy goes down, and normal consumers can buy items for MSRP. If everyone is botting, then no one is botting.'[22] FairGame is a great example of an algorithm that can be owned by individuals. Still, basic software development skills are needed to use it, and for that reason it's not an algorithm that is accessible to all.

When a business or a third party owns the bot, there are three ways they can deploy them: to perform direct sales, to mediate between shoppers and sellers, and to enable other businesses to interact with customers.[23] Let's have a look at each of these three approaches.

In the case of direct sales, businesses use digital minions to sell goods and services to their customers. HP's InstantInk, which I discussed earlier in this chapter, is a sales agent 'deployed' by HP to its customers' homes.

Mediation occurs when software agents are created by a third party to help bring businesses and customers together. Almost like personal shoppers, these digital minions search for a product the customer needs and then order it. There are many scenarios in which such algorithms would make sense: shopping for the best insurance deals or buying coffee beans from local sellers. But their 'killer app' – the situation in which they're most useful – is shopping for products in high demand and low supply. Professional resellers, who buy in-demand products in large quantities only to sell them at a profit,[24] will use such software to fully automate their purchasing process.

Which brings me to the subject of my son's hobby. My son is a young teenager and, like many of his friends, he has a keen eye for style and design. More specifically, he spends a lot of time staying informed about new developments in the sneaker world: collaborations between brands, relaunches of classic old-school designs and other goings on in vibrant 'sneakerhead' culture. The sneakerhead industry is truly mind-boggling, with record-breaking sales and jaw-dropping valuations. Did you know the most expensive used shoes sold at an auction were Nike Air Ship shoes worn by Michael Jordan? They sold for a staggering $1,472,000. Thankfully, my son is happy to just observe: he doesn't buy rare editions. Many sneakerheads see rare trainers as an investment item and are always looking for an opportunity to buy a unique pair. To this end, they may 'hire' bots to purchase shoes. They might even take advantage of digital-minion marketplaces – websites such as BotMart, Cop. Supply or BotBroker – that offer bots for rent, some for as little as a few dollars a day. These websites are like twenty-first-century versions of temp-worker agencies, providing digital minions not human workers.

Stellar AIO is one such digital minion. Stellar AIO is an 'automation checkout software', which can run hundreds of shopping bots that check online stores for various products and buy them immediately when they 'drop'.[25] Stellar's developers claim it can shop at over twenty sites, including Amazon, Target, Walmart, AMD, GameStop, Tesla and Shopify. The 'AIO' in the application's name stands for 'all-in-one', because the bot can do everything a human must do to purchase a product. AIO represents a whole category of bots that offer full automation. Unlike the bot I used to buy face masks, which required me to enter payment information, AIO bots take care of everything.

Sneakerheads are not very secretive about what they do. In fact, many happily share videos showing how they work with AIO bots. One sneakerhead, Botter Boy Nova, is followed by almost two hundred thousand people on YouTube. Here's how he introduces himself on the platform: 'My name is Nova and I'm a sneaker reseller. I demonstrate my hypebeast sneaker reselling operations from top to bottom. I do live cops, unboxings, tutorials, selling, and much more!' He is open about how things sometimes go wrong. His biggest failure? That one time he spent $3000 to hire a large number of bots to 'cart' a hundred pairs of sneakers that 'dropped' that day. That's a cost of $30 per pair, which would have been easily absorbed by his margins. But he only received three pairs of sneakers in the mail, making the cost $1000 per purchased pair, which would be impossible to make back. What happened? On the day of the launch, the retailer – actively trying to stop resellers from buying large quantities of its product – introduced a limit of three items per street address, and most of Nova's orders were cancelled.

My favourite part of almost every 'live cop' video is when Nova's bots enlist Nova to do some of the work. Yes, there is a moment in each video when their roles are reversed. If a bot comes across a captcha – a challenge that tests whether a website user is human, which I'm sure you've come across many times while browsing the internet – it asks Nova to respond. After Nova has clicked on all the hydrants, bridges and smiling dogs, and has performed any other tasks the captcha requires of him, he gives control back to the bots, who finalise the purchase.

In the economy of algorithms, it's not only humans that hire digital minions. Digital minions hire humans too.

Digital minions that mediate interactions between businesses and customers are transforming industries other than retail too. DoNotPay is a 'robot lawyer', a fully automated algorithm that can appeal parking tickets, obtain birth certificates and even annul marriages on behalf of its human users. And, just like Nova's AIO bots, DoNotPay might soon ask humans to step in and perform functions the algorithm can't carry out itself.

In January 2023, DoNotPay's CEO, Joshua Browder, tweeted that his company was willing to pay a volunteer $1 million to appear before the US Supreme Court at an upcoming trial and argue the case by repeating exactly what the robot lawyer would tell them to say. Browder suggested that the robot could communicate with the person in real time via AirPod headphones. 'We have upcoming cases in municipal (traffic) court next month,' Browder noted, 'but the haters will say "traffic court is too simple for GPT". So, we are making this serious offer, contingent on us coming to a formal agreement and all rules being followed.'[26]

The Supreme Court does not allow electronic devices in the courtroom when it is in session, so it's unlikely to hear from an algorithmic lawyer anytime soon. Aside from that, human lawyers seem to be very unhappy about the prospect of an algorithmic defendant. Apparently, Bowder's tweets led to several investigations by US bar associations, whose goal is to uphold the legal profession.

In late January, Bowder posted another tweet:

> Bad news: after receiving threats from State Bar prosecutors, it seems likely they will put me in jail for six months if I follow through with bringing a robot lawyer into a physical courtroom. DoNotPay is postponing our court case and sticking to consumer rights. Specifically, lowering medical bills, cancelling subscriptions, disputing credit reports, among other things, with A.I. I think it's very important for companies to stay focused. Unlike courtroom drama, these types of cases can be handled online, are simple and are underserved.[27]

It seems that courtrooms, at least in the United States, will remain free of digital minions in the foreseeable future. But somehow I feel this is not the end of the story.

Is the world of B2A2C only within the reach of tech-savvy users such as Nova, or Silicon Valley businesses? Thankfully not: that's where enablement – the third category of the B2A2C approach – comes into play. Just as platforms such as eBay created opportunities for almost anyone to sell products online, there are now platforms that allow businesses to deploy digital minions without building the entire infrastructure from scratch. With just a tiny bit of programming, any business can integrate the services of smart assistants such as Amazon's Alexa and enhance them with new 'skills'.

In 2018, Virgin Trains became the first train operator worldwide to sell tickets and provide information through Amazon's Alexa.[28] In just one move, it expanded its sales-office network to include its customers' homes.

As I write this book, OpenAI, the creator of ChatGPT, is testing such integrations with its chatbot. Soon we will all be able to ask ChatGPT to suggest the cheapest way to travel around Europe, and not only will it be able to provide us with ideas, but it will also be capable of booking our flights and train tickets. And it will be easier to communicate with them verbally. While smart speakers can currently respond to precisely formulated commands – they will understand 'Add toilet paper to the shopping list' but might be confused by 'Chuck some dunny rolls on the shopping list, mate' – the next generation of speakers, using platforms such as ChatGPT as enablers, will hopefully be able to understand more colloquial language.[29]

Algorithm whisperers

There is something quite unsettling about most – if not all – of the digital minions I've discussed in this chapter. While the complexity of many modern algorithms makes them 'black boxes' to the average person, these digital minions are even more opaque. As with 'black box' algorithms, we find it difficult or impossible to understand the 'how' of their functionality, something many of us are quite content not to know. But these digital minions also challenge us to ponder the 'why' of their specific decisions – and whether can we have an impact on those decisions.

Let's delve into this a bit further. Remember the ATM that proposed local dining options to my student and other tourists? If the ATM's recommendations genuinely influenced tourists' decisions, the owners of restaurants that weren't recommended would probably be 'intrigued' (to put it mildly) as to why they had been overlooked. Once their initial frustration subsided, they would want to know how to get the ATM to advocate for their restaurant.

Need a different example? I recently introduced ChatGPT to a large group of people who owned private medical practices. Many of them were doctors turned entrepreneurs, often with limited business acumen and even less knowledge about technology. I demonstrated how they could use the chatbot to generate a comparison table of software applications, detailing the pros and cons of each. As it turned out, a software vendor was present in the room and noticed the chatbot did not include his software in the table. It's not hard to imagine the potential negative impact on the vendor's business if ChatGPT continues to 'overlook' its software. If the software wasn't ranking high enough in Google's search results, the vendor would know how to handle it – there are businesses, namely search-engine optimisation (SEO) agencies, whose sole focus is to understand Google's algorithms in order to boost clients' website rankings. But getting ChatGPT to recognise and endorse your product is an entirely new challenge. We're still figuring out how to sway AI agents.

In the software world, there's a term for people who are highly skilled at using computers and applications. These individuals know all the tricks and shortcuts, and they get things done quickly. They are 'power users'. Typically, a power user learns by studying software manuals or methodically going through every available function of an application. A good understanding of an application allows the user to get the most out of it.

The challenge is steeper with digital minions like the ones I've discussed, particularly those governed by advanced AI algorithms. Many of these digital minions don't have clear-cut guides explaining their decision-making mechanisms; and experimenting with them to understand their functionality, as one might do with an email app or text editor, is often not feasible. Things get even more complex when a digital minion acts as a mediator between two parties, as exemplified by the B2A2C scenarios I've described.

People who successfully decipher the complexities of algorithms and discover how to sway their behaviour where possible are akin to algorithm whisperers. You'll encounter a few of these intriguing figures in the next chapter.

But before we move on, let's imagine a world – it's just around the corner – in which our fridges self-replenish, as some of our printers already do.

On the one hand, it would be important for fridge owners to become power users to ensure the fridge's algorithm was *aligned* – that is, doing exactly what it was expected to do, such as consistently finding the best prices for the products it orders.

On the other hand, comprehending the algorithm's decision-making process would be just as important for retailers. Why? Because then retailers might be able to find a way to influence the algorithm's decisions, or they might realise they need to modify their offerings so that the algorithm will purchase from them. Advertisements and product designs that target human consumers use a similar strategy – the only difference in this scenario is that an algorithm is the target audience.

Advertising to a fridge – sounds weird, right? Yet in a world where bots are entrusted with buying and investing, there are numerous businesses that may wonder how to influence the decision-making of digital minions. If it's possible, how will they go about it? Just as behavioural economists try to understand human decision-making in order to influence our behaviour, algorithms are also being analysed.

SEO experts analyse the behaviour of search algorithms, devising ways of getting them to rank some websites higher than others. Other analysts try to decipher how algorithms set prices so they can predict the best time to buy.

If you think back to my anecdote in Chapter 3 about two copies of a reference book being sold on Amazon for eight figures, you might remember that by observing the Amazon bots' behaviour, the genetics professor Michael Eisen was able to determine the rules that governed their pricing, which kept rising. If Eisen had also figured out how to make the bots reduce their prices, he would have successfully influenced their behaviour and, in the process, become the first known algorithm whisperer.

5

Can an Algorithm
Take Your Job?

Leo Tiffish graduated in 2009 with a degree in computer science.★
A 2.3 GPA would make some students question their future, but
it didn't stop Tiffish from looking for a job in Silicon Valley. They
made the right choice. A few months later, they landed a role in a
new Bay Area startup that in the seven years that followed would
become incredibly well-known. But Tiffish's career would take a
turn for the worse, and in 2016, after being fired, they sought advice
on their next steps from users of the website Reddit.[1] Here's part of
the message they posted on the site:

> From around six years ago up until now, I have done
> nothing at work. I am not joking. For 40 hours each
> week I go to work, play *League of Legends* in my office,
> browse Reddit, and do whatever I feel like.
>
> In the past six years I have maybe done 50 hours of
> real work. So basically nothing. And nobody really cared.
> The tests were all running successfully.[2]

Tiffish made a salary of $95,000 per year on average – relatively low
for a computer-science graduate in Silicon Valley, but a very decent

---

★ Leo Tiffish is not a real name – it is an anagram of @filetoffish1066, the
username of the redditor who shared their story. Tiffish didn't reveal much
about themselves, not even their gender, though we know Leo attended a
gym and liked to drink alcohol 'with the boys' – draw your own conclusions.

wage in most parts of the country. Tiffish's job was to test code written by other developers in the startup. It is a common role: a dedicated employee takes software codes – or algorithms – written by in-house developers and tries to 'break' them, looking for errors and other issues the developers might have missed. It is a mundane task requiring much precision, but if done well it is crucial – it helps businesses avoid disasters when they release their software to the public.

Tiffish didn't particularly enjoy the tasks their role involved, but they weren't in a position to complain: they understood how lucky they were to get that job. Instead of looking for another one, they came up with a solution to free themself from the mundanity. They started writing scripts and automating individual tasks. They said it took them about eight months to create algorithms that could do what they were meant to be doing themself. And they likely created them while doing the work they were paid for. They don't seem to have given themself too much credit for this achievement.[3] Ultimately they succeeded: Tiffish automated all of their work. Over the next six years, their responsibilities were carried out by a small army of digital minions.

Robota is an old Slavic word describing mundane, almost slavish work – the type of work Tiffish was hired to do. The Czech playwright Karel Čapek derived the word 'robot' from *robota*. In his 1920 play,[4] *Rossum's Universal Robots*, Čapek uses the word to describe artificially created workers that perform tasks humans don't want to do. We use Čapek's word every day now. When we use it, we usually think of humanoid machines such as the Terminator or *Star Wars'* C-3PO and (a bit less human-like) R2D2. But the term is much broader than that.[5]

Tiffish rarely spoke to anyone in his office. 'I shit you not,' they wrote, 'I had no friends or anything at work either. So nobody ever talked to me, except my boss and occasionally the devs for the software I was testing.' Their manager didn't seem to care too much about checking in with Tiffish, beyond making occasional small talk. This will be a topic of many business-school sessions: how can a manager not notice, in six years, that their employee has fully automated their work?

Tiffish's daily routine was consistent. They did whatever he felt like every day: 'It turned out to be six fucking years of nothing but drinking alcohol, *League of Legends*, some *Counter-Strike* and hitting the gym.' Then, one day, the IT support team looked at Tiffish's computer logs. Their conclusion was obvious: Tiffish was not spending their time testing code. They discovered the programs Tiffish had created, and they notified their boss.

When Tiffish's boss reached out to them, Tiffish admitted they had automated their job. In response, the boss fired Tiffish.

Let's consider the manager's decision for a while. It's unlikely they were prepared for such a situation. Their task was to manage a software engineer responsible for quality assurance, but Tiffish was something else. As a 'people manager', the manager was likely trained to focus on the human aspects of their employees' work, ensuring they show up, do their tasks properly and work well with others. On this basis, they probably thought firing Tiffish was perfectly justified. And of course, they had reason to be concerned: there was no way to ensure the algorithms Tiffish created worked as well as a human should. For instance, what if the algorithms dealt with confidential information in a way that breached company rules?

But ultimately the conclusions the manager drew were wrong. Not only should Tiffish have remained employed by the startup, they should have been promoted. Tiffish had proven their ability to identify tasks that could be automated, they had demonstrated how to automate them and, finally, they had shown they knew how to manage such algorithms in the long term.

But Tiffish's manager did not appreciate the value of this. It was as though their primary goal was to make Tiffish do the mundane work – the *robota*. This fundamental mismatch between the manager's expectations and Tiffish's actions caused the manager to let go one of their most brilliant employees.

In the last few years, I have used the case of Leo Tiffish in my university lectures and during keynotes. The story always captivates my students. Secretly, many of us wish we were like Tiffish. We want to automate the boring, repetitive tasks we're asked to do. We want to spend more time on what fulfils us. But the story is as much about Tiffish's manager as it is about Tiffish. To make that point, I always conclude the story by expressing my belief that firing Tiffish was one of worst management decisions ever.

Digital minions are challenging our perception of what work is. While the business world has used automation technologies for a long time, employees are now proactively automating too. Could this employee-driven automation get out of hand?

What would you do if you found out that one of your friends, colleagues or employees had fully automated their job? I hope that after reading this book you would react with excitement, a tiny bit of caution and a good plan for what to do about it – especially if you're their manager.

When digital minions do our work, it can be tempting to step away – just as it can be tempting to move to the back seat of a self-driving car, leaving the driver's seat empty. But should we? Are digital minions ready to take the wheel? Some organisations have addressed this dilemma by hiring people to manage algorithms and make sure they don't go rogue. Can this be a fulfilling job? And what if algorithms are asked to manage humans? How do people respond to being managed by an algorithm? Let's unpack some of these questions.

Grassroots automation

A couple of years ago, I received an email from a student named Matt. He asked for some career advice. Matt wanted to understand how I managed to balance industry work with academia. I thought this was an excellent opportunity for Amy, a new team member, to find time for Matt and me to meet. I copied Amy into the email trail and asked her to set up the meeting. Sixteen minutes later, Amy replied. She wrote that she was trying to find some time for us to meet on 28 February. She also noted that she couldn't find an empty slot in my calendar during my preferred meeting hours and suggested scheduling something at a different time. Simultaneously, she emailed Matt, asking whether 9.00 a.m. would work. About twenty minutes later, I asked Amy not to schedule anything on 28 February. After another few minutes, Amy emailed Matt, with three alternative options. The following day, Matt replied that the last option would work well for him. Amy acknowledged his response and followed up with a meeting request for the two of us.

The above is a fairly typical example of how much coordination the scheduling of even the most straightforward meeting requires. But there's a twist: Amy was not human. She was an algorithm I hired to act as an agent. The algorithm, also available under the name 'Andrew', was created by New York–based X.ai, a technology company with a mission to 'democratise personal assistants'.[6] Amy is an example of *shadow automation*: my employer doesn't know I use algorithms to schedule meetings. But I am not too secretive about this fact either. I used my corporate credit card to pay Amy $9 a month.[7]

The ability of employees to automate business operations, and not just their own actions, is a recent phenomenon. You once needed an IT department, a bunch of engineers or a close partnership with a vendor to automate tasks. Not anymore. The *grassroots-automation* movement is growing. Al Sweigart's how-to book on the subject, *Automate the Boring Stuff with Python*, has sold over half a million copies, which is incredible, considering the book is freely available online as a PDF.[8] A Reddit forum titled 'Automate' is full of conversations about automating various aspects of work. The second most popular topic in the forum is 'I just automated 100 hrs of labor per week at my company'.[9] An employee of a small business wrote the post. He automated the process of taking orders from customers. 'We are a small company so we didn't have to fire anyone,' he wrote. 'We just re-purposed our CSRs [Customer Service Representatives] towards helping the customer more.' The automation was achieved without an IT department or an external vendor. Sometimes one skilled employee is all it takes.

Grassroots automation could start with the simplest of tasks: using a few keystrokes to write an email. I often receive inquiries from potential PhD students – more students than I can supervise. With six keystrokes – '/nophd' – I can get my computer to write two paragraphs telling the student that I am unavailable and explaining what other options they have. Most responses now take me a few seconds instead of a couple of minutes. Underwhelming? Over time, these saved minutes add up, and I get hours of work back.[10]

The more recent emergence of large language models is making the automation of slightly more complex tasks just as easy. These models, which power chatbots such as ChatGPT, often surprise their creators with so-called emergent behaviours: skills they were not explicitly designed to have. One such skill is writing software scripts. I tested this ability a few days ago, when I asked ChatGPT to write a script that would automatically respond to some of my emails. To my surprise, ChatGPT was able to write code in AppleScript, a programming language that I don't know well, but which is the most straightforward approach to automating tasks on my computer. And the code worked.

On the more sophisticated end of the grassroots-automation spectrum are Leo Tiffish's scripts to automate their work tasks. Custom pieces of code might require significant effort to create. But they save you much more time than you spend building them.

I regularly come across stories similar to Tiffish's, but with happier endings. In one case, a night auditor at a hotel automated most of their jobs and, just as Tiffish did, tried to hide this fact from their employer, fearing they would be fired.[11] After a bit of nudging from 'redditors' – people who read the story on Reddit – the auditor decided to show their digital minions to their manager. The manager was so impressed that they promoted the auditor and arranged to use the algorithms at other branches of the hotel.[12]

Another employee, based in the Netherlands, automated their software development job.[13] Unfortunately, the developer was fired by their manager after they confessed to the automation, but the developer decided to appeal the decision with the manager's boss. When the superior learned how much money the developer's automation had saved the company, they brought the developer back on board and promoted them. They also fired the trigger-happy manager who had fired the developer in the first place.

Some employees are realising the potential presented by grassroots automation. With a small army of digital minions giving them back their time, it makes sense to consider 'reinvesting' it. One option is to do more for their current employer. But why not get another job? And then another one?

In my research, I have come across several cases in which employees claim they receive numerous full-time salaries while delivering all of the outputs their employers expect from them. Is it legal? It depends on their employment contracts. Is it moral? The jury is still out.

The perfect storm of the emerging economy of algorithms and the Covid-19 pandemic, which has afforded people more opportunities to work from home, has given rise to employees like @dreyfan, who shared their story on Y Combinator's website Hacker News:[14]

> I currently have 10 fully remote engineering jobs. The
> bar is so low, oversight is non-existent, and everyone

is so forgiving for underperformance I can coast about 4–8 weeks before a given job fires me. Currently on a $1.5M run rate for comp. this year. And the interviewing process is so much faster today, companies are desperate, it takes me 2–3 hrs of total effort to land a new job, with thousands to choose from.

I was not able to verify @dreyfan's claims, and some of the website's readers thought their post was a joke. *Is it even possible to hold so many jobs at the same time?* they wondered. I would not be so quick to dismiss it. I was once part of a team of remote workers whose workdays overlapped by just an hour or two due to time-zone differences. I was starting my day when other members of my team were about to finish theirs. What I did outside those overlapping hours was entirely up to me, as long as my job got done. Here in Australia, I know many others who work for US- or European-based businesses and enjoy similar freedoms. Perhaps holding down ten positions is excessive, but other than employee contracts, what is to stop people from automating their jobs and being hired by several businesses?

Whether they were joking or not, @dreyfan provided more details about their approach to working multiple jobs. They claimed to work about one hour per job every day: 'I provide enough actual progress that people must assume I'm either extremely methodical or woefully incompetent.'[15] Being seen as available throughout the workday was a plus, they said, so they log into instant-messaging platforms such as Slack and disable the 'idle indicators'. They also use 'virtual machine' software to create 'virtual work laptops' on just one computer. Occasionally, improvisation is needed: 'The biggest problem I run into is video meetings occurring at the same time … I usually just disable video and blame it on internet problems.'

@Dreyfan is blunt about their employers: 'I target overly funded growth-mode companies [which are] focused on adding unnecessary headcount to work on poorly defined projects.' Such companies often have inexperienced leaders who do not know how much to expect from employees or how to manage poor performance.

In the comments beneath @dreyfan's post, others admit to also holding multiple full-time engineering jobs and share their tricks. @Lilbop says: 'I currently run 3 full time contracts concurrently, and yes meetings clash, but it's easy to get meetings rescheduled if you just ask.' @Lilbop adds, 'You'll be surprised how bad you have to be to get fired. I can go days doing nothing, then crank [out] a few hours' work for one pull request on a Friday and people are happy.'

In yet another twist of the economy of algorithms, employees, equipped with digital minions to automate their work, are now exploiting their employers. The asymmetry of power is flipping.

Both employers and employees should beware. Grassroots automation can change from opportunity to challenge if it is hidden from view, becoming shadow automation. Employees who hire or build digital minions and don't disclose it make it impossible for management to plan for the possibility of things going wrong. And if an employee who used shadow automation to do the work of several people were to quit, it would leave their employer short-staffed. Automation can also have negative effects on the employee: after six years of 'doing nothing' at work, Leo Tiffish admitted that their skills in software-quality assurance had deteriorated.

Who is in charge here?

What about automating manual tasks? Here, algorithms that operate inside robots are performing a lot of activities previously carried out by people. I get to interact with one such robot almost every weekend, when Okuri (送り) delivers lunch to our table.

Okuri is a robot waiter working in a shopping-centre restaurant. But do not expect to see a robot such as C-3PO. Okuri doesn't have legs or arms or a shiny shell. Not even a hint of a posh British accent. Okuri consists of a stack of four red food trays and a tiny screen in a black-and-white frame, which sit atop a platform with several wheels. It behaves as though there were a robotic vacuum cleaner underneath it, skilfully navigating around the restaurant.

The restaurant offers a very modern experience. We order the food through an app, and when it is ready Okuri emerges from the kitchen area and brings it to us. The kids love it – the interaction feels very sci-fi to them – but to me it feels sterile. Just like our vacuum cleaner at home, Okuri easily gets stuck and – cue David Attenborough voice – falls prey to a robot's most lethal predators: children.[16] Young humans know Okuri is just a bunch of wires, motors, sensors and chips. I've watched children 'abuse' Okuri more than a few times. I've never witnessed any direct violence, but I've seen children stand in the way of the robot on several occasions, just to see what it would do.

Did I just suggest that children are capable of abusing robots? Very much so. In one experiment, researchers observed a single robot roaming a shopping centre and witnessed children 'persistently obstructing the robot', 'bending the neck' of the robot and hitting the robot with a plastic bottle. The behaviour described by researchers ranged from verbal obstruction to physical destruction. In short, the children were savage. One of the researchers' recommendations was to program robots to escape from children by moving closer to taller people. The researchers specify the 'safety height' as above 1.4 metres.[17]

Admittedly, academics seem to have a strange fascination with the topic of robot abuse. In the paper 'Teaching Robots a Lesson: Determinants of Robot Punishment',[18] a group of researchers investigates why, and when, people decide to punish robots. Other papers explore topics such as 'Is It OK to Bully Automated Cars?'[19] and 'The Morality of Abusing a Robot'.[20]

I am a bit guilty of robot abuse too. I once brought a floor-mopping robot to a panel discussion I was speaking at. It quietly cleaned the floor around us while we discussed the morality of artificial intelligence. But I didn't bring in the mop to keep the area clean. I was more interested in observing people's reactions if I kicked it vigorously whenever it got close to my chair. I saw shocked looks in the audience, but it wasn't until someone shouted, 'Please stop! It's not okay!' that I desisted. A similar experiment inspired another academic paper: 'Kicking a Robot Dog'.[21]

What's with all these academics and their robot-abuse research?

In the twenty-first century, it is increasingly rare for humans to work without the support of algorithms. Likewise, algorithms rarely work without humans. And in mixed human–algorithm teams, it might be unclear who's doing the work and who should be accountable for any issues that arise.

We're often so impressed by the potential of digital minions that we give them a lot of power without ensuring the proper checks and balances are applied, but it is better to assume a clear line of responsibility and have an onboarding process for algorithms, closely supervising their work before letting them work more autonomously. Basically, you should treat them the same way you would treat new human employees. Failing to do this can lead to disaster.

One such disaster happened recently in Australia. In July 2016, the Department of Human Services launched a system to automatically detect any overpayment of welfare to recipients and to calculate their resulting debt.[22] Initially called the Online Compliance Intervention, it was later renamed the Robodebt scheme.[23] Robodebt algorithms compared citizens' welfare payments with tax returns to identify those whom the department believed were liable to return some of the payments they'd received. The system was capable of issuing a whopping 20,000 debt notices per week. Over one million notices were sent out in the system's lifetime. That's a lot. The adult population of Australia is about twenty million.

But there's a difference between quantity and quality. At least 470,000 of the notices issued – almost half of them – were wrong. The Robodebt algorithm asked citizens to return money they didn't owe. Debt notices were issued to deceased people and disability pensioners. Unsurprisingly, people pushed back. A class action against the Commonwealth of Australia led to a settlement of close to A$2 billion in 2021. This is a shocking example of how a rogue algorithm was allowed to continue working for months, delivering incorrect results. A lack of human supervision contributed to one of the biggest scandals in the history of digital minions.

The Robodebt case shows that our legislation is not yet ready for this new division of work between humans and algorithms. You might recall that I believe we should label algorithms according to their level of autonomy, the way we label self-driving cars: from level-one algorithms, which merely support people in their work, all the way to level-five algorithms, which are fully autonomous. Somewhere in between are level-3 algorithms, which require human approval for their work, and level-4 algorithms, which are observed by humans who look out for unexpected behaviour but don't need to explicitly approve each output.

While I haven't been close enough to the Robodebt case to confidently state what problems were involved, I believe a misunderstanding about how capable the algorithms were contributed to the disaster. If the algorithms had been labelled as level 3, it would have been clear that it was necessary for humans to review and approve every one of the algorithms' decisions before their release. On 7 July 2023, a report by the Royal Commission into the Robodebt Scheme was released. It highlighted that the use of income averaging to estimate citizen debt was 'essentially unfair, treating many people as though they had received income at a time when they had not'. The algorithms weren't evil. But they amplified the unfair method at a scale no completely human-run administration ever could.[24]

When I worked with Australian legislators on reforming the administrative law in Australia,[25] my recommendation was, unsurprisingly, to implement a labelling system for administrative decisions. An additional bonus? If you were the subject of such a decision, you could request that the decision be made in a different way, depending on the level of automation that was initially used. For instance, if no human oversight was involved in the decision (say if a level-four algorithm was used), you could ask for the decision to be reviewed by a human. Let's hope that in the near future the public sector will acknowledge the different capabilities some algorithms have and assign them the right level of human support.

There are some uplifting stories about algorithms and humans working together. I find one particularly heartening. It started as an experiment. On 26 November 2018, a pop-up cafe opened for two weeks in Tokyo.[26] It probably won't come as a surprise to hear that it was staffed by robots. You might be imagining robots similar to those at my weekend lunch place, perhaps slightly more android-like. But it wouldn't be Japan if the cafe hadn't come up with something that instantly felt much more futuristic than anything else available to us.

The team behind the cafe recognised that conversing with a robot is not the same as conversing with a human. If a robot recalls our name when we return to a cafe, we might admire the fact that someone programmed it to do so; if a human remembers our name, we know they care. There is a special connection there. We will be happy to come back again.

Can you hire robots to staff your cafe and make it feel human at the same time? The cafe founders' answer was, 'Yes, you can.' Their solution was to hire humans *and* robots.

Avatar Robot Cafe DAWN ver.β initially deployed three OriHime-D robots.[27] OriHime-Ds are not autonomous but are remotely controlled by human 'pilots'. The cafe hired ten pilots: five women and five men. The pilots were disabled, bed-bound and otherwise unable to work or interact with the outside world, certainly not as waitstaff. But thanks to robotic avatars, they could see, hear and interact with cafe customers. It was a wonderful experience for the customers, and the pilots found the work fulfilling and appropriate for their level of disability. 'With the spread of avatar robots and avatar work, it is possible to realize a society where we can do anything if we have free mind, even though we are bedridden,' said the creators of the cafe experiment.[28]

The cafe is now out of the experimental phase. A permanent laboratory cafe opened in June 2021 in the Nihonbashi district of Tokyo. As of early 2023, it has seventy OriHime-D pilots on staff.

The developers of OriHime-D call their machines demi-robots.[29] It's an intriguing idea to bring humans and robots together to perform tasks they couldn't carry out alone. In this case, the robots provide able bodies, and the pilots offer human warmth. The Tokyo experiment shows there is no need for competition between human and machine. It is possible to bring the two together and bring out the best in each. If customers prefer the 'human touch' but a business wants to automate, demi-robots might be the way to go.

May I speak to your manager?

When you interact with a business and something goes wrong, you can always ask to speak to the manager, whose responsibility it is to ensure their employees do a good job by their customers. You'd be right to expect that if a robot waiter at a cafe were to spill coffee on you, a manager would step in at your request – or better yet, without your saying anything – to right the wrong. It is almost unbelievable, therefore, that many digital minions aren't properly supervised and speaking to the manager isn't an option.

The Robodebt scheme is an example of unmanaged algorithms gone rogue. But thankfully more and more organisations understand that algorithms, including those that operate inside robots, need to be managed. And, just like that, these organisations have a new category of employee: algorithm managers.

Nissa Scott started her job in Amazon's New Jersey warehouse in 2016. Her job was hardly a dream: stacking large, often heavy, plastic boxes. Amazon's entry-level warehouse roles are hated by many. Juan Espinoza resigned as a picker – someone who fills plastic boxes with products bought by customers – due to the heavy labour involved: 'We were expected to always pick 400 units within the hour ... I couldn't handle it. I'm a human being, not a robot.'[30] But Scott persevered. A year after joining Amazon, she was given a new task: to be in charge of robots doing the job she used to do.

Amazon, like many other companies, is gradually automating some of the tasks involved in running its business. Its decision to automate its warehouses is helping it to speed up its 'click-to-ship' process. The click-to-ship process dictates what happens between a customer finalising their order online and the ordered items leaving one of Amazon's warehouses. Performed only by humans, the process takes an average of sixty to seventy-five minutes. Introducing robots shortened it to fifteen minutes.[31] For a business that sees next-day deliveries[32] as its competitive advantage, this is a lot of time saved.

In an interview, Scott explained that her new job at Amazon is to 'babysit' several robots. She troubleshoots when necessary – the robots tend to get stuck or confused and need occasional human help. 'For me, it's the most mentally challenging thing we have here. It's not repetitive,' she said.[33]

There are two interesting lessons here.

First, Scott recognises that, contrary to popular belief, working with robots is not a mundane task. It is quite the opposite: all the possibly mundane issues can be predicted and therefore managed without human supervision. It is the non-standard, unpredictable situations that might cause problems – and that's the best time for humans to step in and shine.

Second, one of the most advanced organisations in the world recognises that algorithms need oversight and – in many cases – babysitting. It is the unexperienced organisations that fall for the hype and promise of fully autonomous algorithms and robots. If you see an organisation that claims it has fully automated some of its operations, maintain a safe distance!

Imagine if Amazon's warehouse robots didn't have a person overseeing their performance. It's hard to believe any organisation would allow this. It is perhaps the tangible nature of robots, and our power to imagine what could go wrong, that makes it obvious they need a person to take care of them.[34] The poor beings sometimes struggle in front of our eyes.

But do we feel the urge to help robots reach their goals or are we just trying to avoid disasters? Kacie Kinzer, a student at New York University's Tisch School of the Arts, decided to find out. In 2008, she built several 25-centimetre-tall robots out of cardboard and added a motor, batteries and wheels to make them roll forwards in a straight line. She called her creations Tweenbots. She drew a big smile and large eyes on each of them, and attached a flag with a message on it, asking passers-by to help the Tweenbot get to its destination. Kinzer took the Tweenbots to Washington Square Park, a busy part of New York, less than 2 kilometres from the famous Flatiron Building. Could the Tweenbots make it from the north-east corner of the park all the way to the south-west corner? Or would they get stuck on an uneven surface, fall over a branch or drive onto a road where they would be squashed by a car? 'The Tweenbots' success is dependent [on] people's willingness to step outside of habitual actions and engage with a technological object in the city space,'[35] Kinzer wrote on the project website.

Kinzer recorded a hidden video of Sam, one of the Tweenbots, trying to reach its destination. She was amazed by what she saw. 'Every time a Tweenbot got caught under a park bench, ground futilely against a curb or became trapped in a pothole, some passer-by would always rescue it and send it toward its goal.'

But the people who interacted with Sam didn't just follow the instructions on the robot's flag. Sometimes they made what they thought was a better decision. 'Often, people would ignore the instructions to aim the Tweenbot in the "right" direction if that direction meant sending the robot into a perilous situation,' noted Kinzer. Some people also tried to lecture the robot, even though it was just a cardboard box on motorised wheels. 'One man turned the robot back in the direction from which it had just come, saying out loud to the Tweenbot, "You can't go that way, it's toward the road."'

Can you picture people behaving as kindly towards an algorithm? When software entities make mistakes, they do not receive compassion from us. Possibly because they don't have a physical presence that helps us form an attachment to them, but more likely because we do not know how to help. It is easy for a random person to pick up a Tweenbot and send it in the right direction, but an average computer user has no idea how to help an algorithm that cannot reach its goals. But non-embodied algorithms need help, direction and supervision too. If they perform the same work that human employees otherwise would, we need to ensure they are in the proper 'reporting' structure.

How do you manage algorithms? It depends on who you ask. An organisation's chief technology officer, or CTO, will manage the technical aspects of algorithms. They will use software-management frameworks and policies to manage applications. These tools are essential for maintaining the software, but they don't often address such issues as accountability or impact on organisational culture. This is unfortunate, because when algorithms team up with humans, there might be organisational tensions we have not seen before. Despite this, roles that focus on managing the more 'human' aspects of algorithms are only just starting to emerge.

Recognising that the oversight of algorithms may need to span beyond organisational boundaries, governments are introducing roles to oversee this technology, similar to those that exist for accountants, builders, doctors and so on. In January 2022, the UK government's Competition and Markets Authority created a new job: a Director of Algorithm Assessment and Technology Insight. One of its stated responsibilities was 'to determine how to monitor the compliance of the algorithmic systems of large technology firms'.[36] The oversight of algorithms is at the top of the agenda for this particular job. Many other organisations already recognise the importance of this. In future, there will be more management of algorithms, not less.

When humans are in the role of manager, it might bring out the best in both them and the algorithms. For instance, experts believe that most algorithms lack communication, coordination and adaptation skills.[37] This skill deficiency might be considered a shortcoming if algorithms are left to run by themselves. But it is also an opportunity for humans to step in. Just as Scott did when she helped robots do their job at Amazon. Engineers will point out that modern algorithms are far superior than humans at specific tasks. This is true. But in the messy and often unpredictable world of business, humans can still prove essential, as algorithms are simply not able to adapt to unexpected situations, might struggle with coordinating resources to get the job done and may fail at communicating. For now, bringing humans and algorithms together makes more sense than keeping them separate.

There are many unanswered questions. Does overseeing the performance of algorithms lead to the dehumanisation of work? Are we turning people into machines? What could be more boring than watching an algorithm do its job? Examples such as Nissa Scott's show that it might be possible to design management tasks for humans that are neither dehumanising nor boring.

But let's take it even further. What happens if the roles flip and algorithms become the managers?

The people strike back

One of the managers in a company I worked for, a person responsible for the sales team, once said, 'As of now, all annual reviews will consider only data. There is no need for any conversations.' 'Annual review' is corporate jargon for a conversation with your manager about your performance over the past twelve months. You are expected to discuss such questions as 'Did you meet all the goals you had at the start of the year?' and 'Do you deserve a bonus, pay rise or promotion?' According to our sales manager, these questions could be answered by an algorithm. I remember cheekily asking myself, *If the manager sees themselves as a data processor, do we even need them? Should we replace them with an algorithm?*

I worked in the research division of that business. Our role was to imagine, prototype and recommend new solutions to business problems – a perfect alignment! While this might sound like a joke, it was within my job description to consider whether it was possible to replace the manager with an algorithm.

Next week, in a playful mood, a few of my colleagues and I built a robot. Not a complete robot, mind you. We focused on what mattered. We built a robotic claw and put a plush toy in its grip. The toy represented a salesperson in our organisation. Then, we connected the claw to the sales database of a fictitious company we had set up to test our research prototypes. We turned on the robotic claw. Whenever sales data was behind the expected target for the plush employee, the claw would tighten. When the results were good, it would let the doll breathe. No empathy, no consideration of factors outside of the algorithm's input. The robot considered only data. Is this the kind of manager we want? Could a robotic claw manage a team well?

Building the robotic claw helped us start a conversation about the issues we might face when digital minions rise to the rank of manager. Thankfully, the actual manager realised his inhumane approach to annual performance reviews needed to be changed.

However, since the time we built the claw boss, algorithmic managers have proliferated. Increasing numbers of employees are being managed by algorithms. And some of them must feel just like a plush toy being squeezed by a cold, robotic claw.

'Algorithmic managers' might sound like entities straight from a futuristic, dystopian novel, but they are already here. Ridesharing platforms such as Uber and Lyft use algorithms to allocate jobs to drivers. No human is involved in making these individual decisions. The algorithms decide who gets a job based not only on information such as the driver's location but also on their previous performance, including the ratings given to them by passengers. An algorithm might decide to stop allocating rides to drivers with a low score – a de facto 'firing' of a driver.[38]

The platforms via which many individuals receive all types of gig work are rife with algorithmic managers. Most workers wouldn't even know they have a manager – human or algorithmic. Still, these managers assign jobs, evaluate workers' performances and provide feedback. This is how platforms can manage millions of workers – drivers, couriers, dog sitters and so on – with only thousands of employees running the platform and overseeing them.

Outside of the gig economy, algorithmic management is a bit different: managers tend to be supported in their duties by algorithms rather than being absolutely automated. For instance, some algorithms help managers sift through the CVs of job candidates.

Of course, as with any automation, there can be issues. In 2015 it was reported that an AI hiring algorithm used by Amazon had discriminated against female applicants: it had been trained using past application data, which reflected that previously it was mostly men who had applied and been hired. The algorithm learned to repeat this tendency. When any references to gender were removed from CVs, the algorithm still found a way to identify female candidates. Over the next two years, Amazon tried to make the algorithm 'behave'. But the algorithm kept finding new ways to discriminate against women. In 2017, executives shut down the project.[39] If these potential biases are managed, however, an algorithm such as this could be a great tool for helping managers to quickly comprehend the large numbers of documents they've received.

There are also algorithms that help managers keep an eye on their employees, but, once again, they need to be deployed with caution. During Covid-19 lockdowns, many businesses began to monitor their employees. Their approaches varied, but some of their solutions included tracking mouse movements, keyboard use or the time employees spent using key company applications. Unfortunately, such surveillance had a negative effect on job satisfaction and increased employee stress levels.[40] Conversely, using algorithms to provide feedback on employees' performances seems to increase employee engagement: algorithms are perceived as more fair than humans.[41] In other words, we do not like it when algorithms spy on us, but we don't mind it when they give us feedback on our work.

What exactly makes human hate algorithmic managers? Research identifies constant surveillance, little transparency and dehumanisation as the main factors.[42]

People don't enjoy being closely observed: 'You're being tracked by a computer the entire time you're there. You don't get reported or written up by managers. You get written up by an algorithm,'[43] complained one Amazon warehouse employee.

We also don't like it if we don't understand the reasons behind specific decisions, and algorithms can be opaque. When Ellie O'Byrne, a writer for the *Irish Examiner*, tried to find out how much Deliveroo riders make, she could not figure it out – even when she became one herself.[44] Delivery riders are paid 'a variable overall delivery fee, at a minimum of €4.30, calculated algorithmically based on a distance and time calculation', she reported, quoting the company. 'Riders don't have access to any other information than this on why they are offered a certain fee for a delivery and can't see where they'll deliver to until after they accept the order.'

Finally, people don't like it when algorithmic managers don't treat them like humans. 'I felt like a scurrying automaton after just three hours, at the beck and call of an artificial intelligence that could send me anywhere it wanted,' wrote O'Byrne of her experience at Deliveroo.

But people are clever. If they can figure out how an algorithm works, they can exploit its weaknesses.

Uber drivers in Washington, D.C., did exactly that, and shared their tactics for outsmarting their algorithmic managers with *ABC7 News*.[45] 'All the airplanes, we know when they land. So, five minutes before, we turn all our apps off, all of us at the same time ... [The fares] surge, $10, $12, sometimes $19. Then we turn our app on. Everyone will get the surge,' said one driver. The group informally calls itself Surge Club.[46] Its *algorithm resistance* confuses the Uber algorithm responsible for assigning jobs. It assumes the supply of drivers at the airport is too low to meet demand. In response, it turns on surge pricing to attract more drivers to the area. But there is a twist. Travellers are not forced to accept surge pricing. In a competitive market, they can try to find alternative transport, including good old taxis. So the Surge Club drivers risk losing customers.

It gets even more interesting. Across the world, in Perth, on the west coast of Australia, a frustrated Uber driver shared his view that taxi drivers are often behind surges. Hoping to capture more passengers, taxi drivers turn on the Uber app while they're waiting at locations such as airports to convince the Uber algorithm there is a high demand for rides.[47] When Uber prices surge, passengers are more likely to take taxis.

It is only a matter of time until people and organisations begin advising others on how to work with algorithmic managers. Their early advice might be simple – for instance, it might consist of applying lessons learned from Uber drivers in Washington, D.C., and taxi drivers in Perth. And then, the 'algorithmic manager optimisation' industry might grow further. Many of us will work with multiple algorithmic managers and will need support to optimise these relationships. Ironically, human employees will need to bring in algorithms to help tame and coordinate algorithmic managers.

What about businesses that simply do not mind if their employees are unhappy about being managed by algorithms? It seems many gig-economy corporations assume they will always find people willing to work under such conditions. But while onboarding new workers using an algorithm might be quick in some industries, there are cases where it takes time, and onboarding costs are therefore high. It might be cheaper to make sure current employees are happy than to find and train new people. Understanding and preventing the potential negative impacts of algorithmic management is usually a rational business decision.

And how do you determine whether an algorithmic manager is harmful to employees? It is not as easy as asking managers or employees – we're often unaware of hidden biases or issues affecting the behaviour of algorithms. But thankfully there are organisations that focus on identifying harmful algorithms. The Algorithmic Justice League, for instance, can help you understand the impact of surveillance and dehumanisation on your employees.[48]

The economy of algorithms is a transformational time for organisations. Digital minions are a new type of employee. Some of them are joining organisations officially, with corporations consciously introducing algorithmic workers and managers, and some of them are joining unofficially, through shadow automation, thanks to employees who are fed up with repetitive and mundane tasks and don't want to wait for the businesses they work for to transform.

These algorithms are not just tools anymore, the way computer software used to be. They are taking over many of the tasks that employees previously performed, and they often do it so well that some of us – customers, other employees, managers – don't even realise they're not human. Digital minions are making us question all of the assumptions about the workplace we might once have had.

While there are some cases in which work is completely automated, this approach usually causes more problems, on a bigger scale, than it solves. We mustn't assume we live in a binary world where a job is either done by a human or done by a robot and human replacement is the norm. There are many reasons for introducing mixed human–algorithm teams. And efficiency is only one of them.

It is refreshing to observe that more organisations are recognising the need to manage their algorithms. They are starting to think of their algorithms as employees, supporting them when they get stuck, and always being ready to correct their actions if they unconsciously make mistakes.[49]

When I work with businesses looking to introduce algorithms to perform traditionally human tasks, I encourage them to think about algorithms differently. What if these algorithms were regular employees? What team would you put them on? Who would be their colleagues and managers? Does diversity – bringing humans and machines together – matter? How would you hire and onboard them? When would you fire or reassign them to other tasks? What might happen if the algorithms took on managerial roles? How can you ensure they perform their tasks properly? Who manages the algorithmic managers? And how will you ensure that algorithmic management does not adversely impact human employees? These are not hypothetical questions. As we have seen in this chapter, these are the fundamental questions that today's businesses need to answer.

6

Do Algorithms Dream of Electric Sheep?

Even the most innovative organisations struggle to reinvent themselves – anything more than tinkering around the edges of an existing business model can seem infinitely complicated. But in order to survive they sometimes have to.

Nokia is one of the few companies in the world to demonstrate both impressive flexibility in innovation and an almost shocking inability to adapt to the changing world. The Finnish company started out in 1865 as a pulp mill. In the first hundred years of its existence, it created products as diverse as toilet paper, gumboots, gas masks and military radios. It wasn't until the 1980s that Nokia expanded into the digital technology space, building personal computers. It entered the mobile phone market in 1987, when it released the Mobira Cityman 900, one of the first handheld mobile phones in the world.

The following twenty years were a dream ride for Nokia. The first GSM (Global System for Mobile Communication) call was made in 1991 on Nokia equipment. In 1998, Nokia became the bestselling mobile-phone brand globally. The first camera phone, released in 2003, was a Nokia. The company became the poster child of innovation and digital transformation. On 12 November 2007, *Forbes* magazine put Nokia's CEO on its front page. The pull quote asked, 'One billion customers – can anyone catch the cell-phone king?'

And then everything came crashing down. There was almost nothing to suggest that Nokia's growth would slow. But as we all know now, the king slayer was emerging. Earlier that year, Apple, another business trying to reimagine its future, announced the first-generation iPhone. To many, the new phone felt quirky – more of a toy for computer geeks than a *real* phone. But it was the start of Nokia's downfall. To this day, Nokia is looking for another opportunity to reinvent itself.

Why is innovation so hard? Because emerging trends – whether technologies or business models – are often considered *not good enough* in mature markets. Take electric vehicles as an example. Ten years ago, owning an electric vehicle was associated with range anxiety: *Will I reach my destination without the battery going flat?* drivers wondered. It caused charging woes. I remember driving a small EV when I lived in Singapore in 2012. I couldn't charge the car overnight where I lived, and I knew of only one charging station on the island, far from my place. The small car, while fun, was a burden. Most car manufacturers did not want to subject their customers to such experiences.

Even though EV technology showed long-term promise, the short-term risks and potential losses stopped manufacturers from investing in it. This left the market open for manufacturers focusing solely on EVs, such as Tesla, which didn't have to worry about inconveniencing their existing customers, because initially they didn't have any. This problem – *Should we invest in a trend that will likely disrupt our industry in the future but make our customers unhappy now?* – is known as the innovator's dilemma.[1]

Paradoxically, listening to their customers made Nokia blind to the potential of the iPhone. Why would they inconvenience customers with a phone that didn't have a keyboard and could barely last a day on a full charge? And yet market disruptors become successful by introducing products that are initially inferior, bring lower profits or address insignificant markets.[2]

The innovator's dilemma is a form of sunk-cost fallacy: a tendency to stick to an established strategy, even when it becomes clear that a change in course might be beneficial. Sunk-cost fallacy is a result of the very human inclination to become emotionally attached to past decisions. What if we could free ourselves from such attachments? Would that make it possible to overcome the innovator's dilemma? This is where algorithms could play an important role: unless explicitly programmed to prefer past decisions, algorithms will unemotionally pursue the directions that are the most promising. Algorithms could be the key to unbiased innovation in business.

Another challenge that innovators face is making innovation boringly normal – it should be business as usual: part of the organisational fabric, rather than a special project. Can the economy of algorithms help businesses that struggle to reinvent themselves? Could AI systems, such as the one I used in my innovation-leadership training, help businesses transform? The short answer is, 'We don't know yet'. But there is a longer, more promising one. The economy of algorithms makes it possible to embed innovation in every organisational activity and make it a routine aspect of doing business rather than a one-off, separate action.

Such algorithmic innovation differs from what we typically picture when we think about innovation. Algorithms do not run brainstorming sessions or design jams – activities that innovation professionals would organise. They don't run customer-empathy interviews, as a design expert would. Instead, they ... mutate – just like organisms in nature, adapting to their environment.

Not every algorithm mutates, mind you. Most algorithms do the opposite: they stay the same, never changing and never adapting unless explicitly programmed to. But there are types of algorithms that are designed to evolve. They can learn from the results of their actions. And occasionally they can introduce random variations in their behaviour. They can then assess the impact of such variations, and if the effect is positive they preserve the changes. These mutations are like micro-experiments replacing much larger 'innovation projects'. And – here's the best part: they can be run automatically and at scale, for activities that are already automated within a business.

Mark Zuckerberg once said that Facebook conducts tens of thousands of experiments at a time.[3] As far back as 2008, when Google was a smaller company than it is today, they used their search engine to run fifty to two hundred experiments at once for any given user visiting their webpage. This scale of experimentation wouldn't be possible without automating the process – from the emergence of a hypothesis to deciding whether to permanently implement a change in the case of a successful experiment. Mutating algorithms make truly continuous innovation possible.

Innovation is a product of curiosity. Walt Disney, a great innovator, is often quoted as saying 'We don't look backwards for very long.

We keep moving forward, opening up new doors and doing new things, because we're curious ... and curiosity keeps leading us down new paths.' Is algorithmic curiosity possible? Could algorithms be creative or innovative?

I feel a slight uneasiness when bringing up topics such as innovation, creativity or curiosity in relation to algorithms. Just a few years ago, I would have declared, very confidently, that algorithms cannot really inspire us beyond using some basic tricks that fall under the 'mutations' category. But now I'm not so sure.

There is a family game called Story Cubes. It consists of nine six-sided dice, with symbols on each side, where normal dice would have numbers. The object of the game is to throw all nine dice and use the symbols that face upwards as prompts to create a story, starting with the words 'Once upon a time ...' It's surprising how creative you can be when you are shown nine random symbols. In the past, I would put algorithms and Story Cubes in the same category of innovation, prompting new ideas with their random suggestions.

But the way algorithms have developed over the last couple of years has been truly mind-boggling. Algorithms such as GPT-4, which entered the public domain in 2021 as GPT-3, can now create documents that are good enough to earn academic credit at universities.[4] In early 2023, when Microsoft integrated a version of ChatGPT into Bing, its search engine, it added a slider for users to choose their preferred 'conversation style'. The three options it provides are: precise, balanced and ... creative.

In 2022, we saw a wave of algorithms able to generate not only text but also visual and audio content that inspires humans. In August of that year, Jason Allen won first place in the digital arts category of the Colorado State Fair Fine Art Competition.[5] His submission, *Théâtre D'opéra Spatial*, was an image generated by an algorithm called Midjourney in response to a carefully prepared description (a so-called prompt) that likely wasn't longer than a sentence or two. And it is truly stunning: definitely not something we would have expected an algorithm to generate prior to 2022.

So far in this book you have met one category of algorithms that challenges our perception of the capabilities of computers: large language models. Large language models are very capable when it comes to everything text-related: they can understand written text and *create* text, including poems, letters and business plans, that reads as surprisingly human. We're still learning about the emergent capabilities of large language models: they can do many things they were not explicitly designed or programmed to do.

But large language models are just one category in a bigger group of algorithms that is called generative artificial intelligence. The Google-made algorithm that composes music, which I discussed in Chapter 1, is another example of a generative-AI algorithm. And Midjourney, the algorithm used by Allen to generate his award-winning image, is another.

The art community is unhappy about the emergence of these algorithms. And this is putting it mildly. 'This sucks for the exact same reason we don't let robots participate in the Olympics,' wrote a Twitter user in response to the news of Allen's win.[6] The Olympic Games are a celebration of human athleticism. We applaud runners in the 10,000 metres category, but during our daily commutes we prefer to use technology to cover such distances. Perhaps it is the same with art, and we should have a 'human arts' category, where we celebrate human ingenuity, while allowing algorithms to roam freely in other categories?

What's the link between business innovation and playing Story Cubes, ChatGPT writing poems and Midjourney creating award-winning images? The generative AI that is currently transforming the art world has the potential to transform the business world too. And the more automated a business is, the more opportunities there are for running automated experiments at scale.

How far could we take business automation, though? In this chapter, I will explore organisations that have gone so far as to be entirely humanless: no human – not even a CEO – is required for them to exist and operate. Such organisations could be designed to evolve without human intervention. It would be a true digital evolution – the survival of the fittest algorithms. At least in theory.

But before we talk about curious and evolving algorithms, we need to talk about us – the curious humans – and what drives business innovation.

Becoming a flâneur

It was a scorching summer morning in Brisbane, and my car's air conditioner was barely managing to cool the air inside. All of the buildings around us looked the same: large blocks of concrete and steel. We finally reached our destination, and I parked the car in the only shaded area around. It was eerily quiet: everyone was hiding indoors, away from the heat. By the time my colleague Friedrich and I made it from the car to the building's entry, fifty steps at most, our shirts were wet. Friedrich, who is usually based in Germany, had joined my team for a few months to research digital transformation in Australia.

Once inside the building, we were met with a wave of cool air and the most cordial greeting: 'Gentlemen, welcome to Watkins Steel!' It was Des Watkins, the managing director of Watkins Steel, a family business that has offered steel-fabrication services since 1968. I pulled out my phone, attached a small microphone and hit record.

Five years earlier, I would have immediately said no if you had asked me to visit a steel-fabrication business. There would have been no doubt in my mind: it was the most uninteresting industry ever. Don't get me wrong: steel fabrication is essential. Without it, the entire construction industry couldn't exist. But was I going to get excited about cutting, grinding and welding? Watching grass grow sounded more thrilling.

And yet there we were, visiting the business and interviewing the team. In fact, I had insisted on going to Watkins Steel. I couldn't wait to see it. Not only because they have drones, lidars and robots that cut steel with hot plasma, or because they use augmented reality in their work and can simulate construction environments to optimise the building process. I wanted to visit Watkins Steel because they had masterfully transformed themselves from a traditional business into a leader in digital services, without losing any of their original strengths. And I wanted to understand how they did it.

I first met Des Watkins in early 2018. The two of us were invited to speak on a panel during an industry event. I am used to having conversations with business leaders at such events. CEOs and managing directors often try to explain why digital transformation is challenging for their business. You know: 'We don't have the right skills', 'Our suppliers aren't cooperating', 'Our customers don't need us to digitise' or 'We're too busy trying to survive'.

But my conversation with Watkins was different. When I tried to ease into the exchange, saying that sometimes you don't need to digitise everything all at once, he disagreed: 'Marek, you do, my friend.' And while I was trying to scramble back into position ('Of course, if you can devote time and people to the task, you should do it.'), he just kept delivering punches: 'We fully automated our production line. We use augmented reality. And we built a hard-hat version of HoloLens [an augmented-reality headset]. Do you think Microsoft would be interested?'.[7] This was intense in a very good way. The exchange felt like it could be happening on stage at some 'SteelFabTech' event in Silicon Valley. I was hooked.

And Watkins was definitely on the right track. In 2021, Watkins Steel received the Optus Enterprise Platinum Award, a winner-across-all-categories prize, at Brisbane's prestigious Lord Mayor's Business Awards. I am convinced they are only just starting.

On that hot morning in 2018, Friedrich and I spent several hours with the team. We wanted to understand how a small-to-medium enterprise (SME) such as Watkins Steel could be so good at digital transformation. The company is an example of masterful curiosity: it has explored opportunities in a seemingly random way to identify the best ways to provide more value to its customers and the construction ecosystem.

Since 2014, when Watkins Steel began its transformation process, it has robotised its production line and started using laser scanners, augmented-reality goggles and digital-design software. Using so many technologies allowed it to start delivering services that had nothing to do with steel: volumetric analysis and construction-site simulations. Realising that it now offered services outside the steel-fabrication industry, Watkins Steel launched a sister business, Holovision 3D.

What Friedrich and I learned was eye-opening. The entire organisation was set up for trend exploration. Even though it was a small business, its employees were encouraged to experiment and embrace uncertainty. They regularly attended trade fairs and events in other industries to get an 'outside-in' perspective. Individual employees were free to purchase technologies and experiment with them, even if it was unclear how these investments would create value.

Des Watkins seems to have found a perfect balance between delivering value to customers and exploring seemingly unrelated opportunities. When I think about the type of exploration he encourages in his business, *flâneur* is the word that comes to my mind. *Flâneur* is a French word describing a deliberately aimless pedestrian.[8] Picture a small French town, teeming with life, its narrow streets filled with corner stores and restaurants. If you ever found yourself in a place like that and decided to walk around and follow your impulses, with no plan beyond just wanting to explore, you were a *flâneur*.

I believe this word can also describe organisations: those that seek to be more curious and creative often engage similarly in deliberately aimless exploration, asking questions and exploring unknowns. Could an organisation fall in love with exploration? Whenever I visit Watkins Steel, and I do regularly, I have no doubt it could. Research shows that curiosity in a business is essential to its success.[9] And *flânerie*, the act in which *flâneurs* engage, is now seen as a valid process for exploring both physical and digital worlds.[10]

It is surprising how many businesses continue to ask the same questions and struggle to find new answers. But when a company behaves like a *flâneur*, it looks for questions it has not asked before. And because so few organisations engage in this process, it's possible those questions haven't been asked before at all – by anyone. *Flânerie* might lead to entirely new solutions to business problems.

Remember the Japanese coffee shop staffed by human–robot teams, which I discussed in Chapter 5? Its founders, like true *flâneurs*, were driven by curiosity and an impulse to explore the unknown. They were putting questions before answers. They referred to their initial experiments as hypotheses. The hypothesis approach meant they embarked on ongoing reflection. They never stop questioning or experimenting, and curiosity continues to drive their business.

A criticism I sometimes hear when I talk about aimless exploration in business is that certain companies don't have the tolerance for failed experiments. In most businesses I know, 'aimless exploration' is not a term you'd use when asking for an innovation budget. These businesses try to increase the success rate of innovation efforts, but their mindset leads to the pursuit of mainly 'safe' experiments, where the outcomes can be predicted in advance. And isn't aimless exploration the opposite?

Businesses need to see things from a different perspective. They should focus not on the success rate of individual experiments but on the absolute number of successful experiments. If a business runs only a few experiments within its innovation budget, it's tricky to maintain a high success rate. But what if it ran many more experiments with the same budget? These experiments would have to be much cheaper and, therefore, smaller. Many promising directions could be found this way. And with a bit of a budget left aside, the most promising areas could be explored further. Curiosity in business is a numbers game. If you run ten thousand experiments and 99 per cent of them fail, a hundred will be successful. And when the services of a business are fully automated, it is possible to run ten thousand or more experiments every day, as Facebook and Google do.

And the hard-hat version of Microsoft HoloLens that Des Watkins mentioned? Microsoft was indeed interested in such a prospect; they just didn't know Watkins had already created it. Later that year, Microsoft partnered with Trimble, a Sunnyvale-based business, to imagine and build Trimble XR10: its own hard hat for HoloLens.

Mutant algorithms

Kartell is an Italian family-business that makes furniture. Like a true *flâneur*, it is fuelled by 'the desire to explore new paths and engage in new experiences dominated by an intense passion for creation', as their website puts it. Founded in 1949, the business works closely with Philippe Starck, a French designer.

Recently, Kartell, Starck and the 3D-software company Autodesk partnered to design and manufacture a new addition to Kartell's collection. They decided to tap into the potential of artificial intelligence to create a unique chair. Autodesk equipped one of its applications with an algorithm to generate designs in response to high-level descriptions, or design requirements. Starck described the design's properties to the AI algorithm, providing specific design criteria and constraints. The descriptions included stability and style requirements: the algorithm was expected to design a chair that was in keeping with Kartell's style and wouldn't fall over. Another constraint nudged the algorithm to use as little material as possible. 'Kartell Loves the Planet' is the company's motto.

The process used to design the chair is called generative design. Generative design involves a close collaboration between a person and an algorithm.[11] First, the algorithm produces a set of candidate designs that meet the requirements it has been provided with.

Then the human designer provides feedback, accepting some designs and rejecting others.[12] These early designs might trigger the human designer to include additional constraints, correcting for the tendency of people to make implicit assumptions that algorithms need to hear explicitly.[13] Based on the initial feedback, the algorithm creates a second 'generation' of designs, introducing variations to the preferred designs from the first round. These variations are the equivalent of mutations in evolution – hopefully generating designs that are more attractive to the human designer. And, just as they did in the first round, the human designer will accept some designs and rejects others. After a few such interactions, the final design emerges.

Kartell published a video showing how the design evolved. It is mesmerising. Initially clunky, like something from the Roblox game, or perhaps like something built with a few Duplo bricks, the chair gradually gets less bulky, its lines soften, its legs straighten and an unusual support structures emerges. In the last seconds of the video, the final design of the chair, called (rather dully) 'A.I.', appears.

Starck compared the process of creating the chair to a conversation:

> Philippe Starck, Autodesk and Kartell asked Artificial Intelligence a question: 'A.I. can you carry our body with the least amount of material possible?'
>
> A.I., without culture, without memory, without influence, replied simply with intelligence, its 'artificial' intelligence.[14]

Kartell started selling the chair in the summer of 2019.[15] Some commenters call it 'surprisingly organic'.[16] Those of us who grew up watching movies such as Tron expect computer-generated creations to be made up of straight lines and right angles. But should organic-looking designs surprise us? The designs are the outcome of processes inspired by nature – neural networks and evolution. It would be more surprising if they did not have an organic feel.

Algorithms can offer similar help to organisations who want to develop other new products and services, reach new customers and get existing customers to buy more. If an 'end goal' is stated in a way algorithms can understand, they can introduce slight variations in existing products, services and processes (such as communications with customers) to get closer to the end goal. By introducing minor changes – mutations – they generate options that can be tested 'live' or reviewed by humans, whether customers or employees. Some get to 'survive', because the customers or human designers like them or because the algorithm itself decides the change is better.[17] Kartell's A.I. chair design underwent hundreds, if not thousands, of mutations before reaching its final form.

Those of us who have spent many years working with artificial intelligence and very advanced algorithms are currently mesmerised by some of the algorithms that are emerging. I run a 'reading club' at my university, in which every week a dozen or so AI academics come together to discuss some of the most recent advances in artificial intelligence. Our discussions are triggered by a weekly read: an academic paper, suggested by one of the club members, discussing a recent breakthrough. And in every meeting, without exception, one of us brings up an AI development to which most of us react with awe. Our minds are blown, because we know how hard, how impossible, it would have been for some of these developments to occur only a year or two ago.

Some of these developments are easy to demonstrate to others. Text-generation or image-generation systems such as ChatGPT and Midjourney are available for anyone to use. I tend to get very excited when I talk about these tools, and I look for opportunities to use them – to help me write a grant proposal, run an ideation session or perform tasks normally done by research assistants. But when I demonstrate the outputs – entire paragraphs written by ChatGPT, beautiful images created by Midjourney – typical reactions include, 'That's nice, but there's a typo there,' or 'Why does the cloud look so strange?' People very easily accept the 'magic' of technology as something absolutely normal. But perhaps I shouldn't be too surprised by this: it's a standard pattern in the progress of technology. It's easy to treat a smartphone as an everyday tool now, but only twenty years ago every single aspect of a current smartphone would have been perceived as mind-blowing.

Still, there is one other aspect of modern algorithms that makes them 'normal' in the eyes of most people – normal as opposed to 'alien' or 'strangely familiar'. Many AI algorithms that have emerged recently were built to create outputs that seem just that: perfectly 'normal'. Something that could be created by people or in nature. These outputs don't feel synthetic or machine-like.

A type of machine-learning framework that has recently proliferated in AI, called Generative Adversarial Networks, pits two algorithms against each other. One evolves its creations, while the other gives feedback on these creations. Imagine one algorithm designing a car, and the other responding, 'Yes, this looks more like a car than your previous design.' (Or, 'This looks less like a car now.') Such feedback helps the first algorithm decide whether it is evolving in the right direction or not. This simple approach has led to the creation of remarkably good algorithms.

They are so good that many people are concerned. When the outputs of an algorithm feel 'normal' and we are not aware they're the creation of a machine, we might fall for them. And when we fall for them, we might attach human-like attributes to them – for instance, when algorithms have produced unintended outputs, people have said the algorithms are hallucinating,[18] and when algorithms have created content that expresses feelings, people have questioned if the algorithms are sentient.[19]

There are many ways such algorithms could be used to mislead humans. You have no doubt heard about deepfakes: video or audio documents that show well-known individuals – presidents, actors, CEOs – doing or saying things they never did or said. The algorithms that make deepfakes can be used to create and spread misinformation if their outputs are of sufficiently high quality to pass as authentic.

In higher education, a particular concern at the moment is the use of text-generating algorithms by students to produce assignments. Because of their capability to 'parrot' text written by humans, these algorithms can generate documents most students are capable of writing: not outstanding, but legible and sometimes way above average. Students can generate entire essays within minutes and edit them to make sure they include required information. No anti-plagiarism software can currently detect how these documents were created. This poses a completely new set of challenges for educators. But an important question that not enough people ask is the following: Could it be okay for students to use such tools? And if so, should we show students how to use these tools to express their own original ideas? Some university educators are now expecting generative AI tools to be used by students, rather than forbidding their use.

Algorithms can help organisations expand innovation, taking it from conscious design to 'unbiased mutation'. Unlike humans, algorithms do not have a preference for exploring particular directions, unless they've been programmed to. Such exploration might randomly go in as many directions as possible. Doesn't it sound like the deliberately aimless exploration of a *flâneur*?

There are AI researchers who explore algorithmic curiosity. They define curiosity in an algorithm as an 'error in an agent's ability to predict the consequence of its own actions'. The worse the error – or inability to predict what will happen – the more curious, and therefore exploratory, an algorithm will be. Like a digital *flâneur*.

In my day-to-day work, I collaborate with organisations to help them shape their digital strategies. Recently, I was involved in a project with a large government department. A particular aspect of our work is that we use a 'futures thinking' approach: we depict potential futures – ten or twenty years from now – and test potential strategic directions against these futures. It is a pretty formalised process. We don't exactly 'imagine' them. Rather, we run detailed background research, identify key dimensions of change and then explore extreme ends of those dimensions. The outcome is a set of very distinct, slightly extreme futures. We present these futures, which we call 'scenarios', to various research participants to get their perspective on strategies we develop to address them.

For this project, we needed to generate four scenarios for the Australian state of Queensland in 2030. We had completed our background research, identified key dimensions of change and defined the high-level ideas about each of the four scenarios – we just had to write four stories that neatly summarised the research. In slightly simpler language: we had the digital equivalent of a few hundred Post-it notes that had to be turned into four scenarios.

Normally, we'd have to spend a couple of hours writing descriptions of each scenario. Often it would take much more time than that. To write good scenarios, you have to carefully consider every insight in your research and decide whether to include it or not. Then, you have to write a description that people will read. It needs to flow, make sense and tell a story. It takes lots of work.

In our projects, we have a rule: even if the task is repetitive and we could simply follow a process, we always introduce experiments into our work.[20] This time we decided to use an AI algorithm to generate scenario descriptions.

I used my very basic software-development skills to write a simple script – a digital minion, if you will – with a specific prompt and a few nerdy parameters, which would connect to OpenAI's GPT-3 servers.[21] The prompt asked GPT-3 to generate a description of the future. I also provided the algorithm with our research (the Post-it notes).

What GPT-3 came back with was quite impressive. Here is the text generated for one of the four scenarios. Have a read. There's a typo or two, but other than that we could use it straight-away! It's pretty mind-blowing when you consider it was written by an algorithm.

> Despite a strong focus on digital transformation in environmental regulation, Queensland is facing a number of unmanaged environmental disasters in 2030. These disasters are having a significant impact on the environment and the economy, and the government is struggling to cope with the situation.
>
> There has been a great deal of investment in sensor and surveillance technology, and algorithms are used to predict and detect noncompliance. However, this has not been enough to prevent a number of catastrophic disasters.
>
> Industry is calling for extended timelines for climate action, and there is a great deal of scientific pushback against the environmental impacts of industry. The trust in environmental regulators is low, and there is a feeling that they are to blame for the disasters that have occurred.
>
> Digital transformation has faced a number of challenges, including delays, budget blowouts, and incomplete systems and services. As a result, the monitoring and mitigation of natural hazards has been less successful than hoped.

With this new approach, we generated each scenario in less than ten seconds. I am now able to do a full day's worth of creative work in less than a minute.

Such digital evolution is a novel way of innovating in business. And several industries are already mastering this space. Aeroplane manufacturers use generative design to create plane parts that weigh less. Automakers develop vehicle components that are sturdier yet lighter. Artists use AI algorithms in their creative processes. No matter which industry you are in, it has either already been affected by digital evolution or it soon will be.

Humanless organisations

How far could we take automation? Could new technology allow us to create previously impossible types of organisations?

These questions were on the minds of Stephan Tual and brothers Christoph and Simon Jentzsch in early 2016. At that time, a new technology, blockchain, was gaining attention. A blockchain is a digital record-keeping system for various types of transactions that securely stores information across a network of computers. Every time a new transaction is entered, it creates a 'block' that is attached to the previous transactions, forming a 'chain', hence the name. This linking of transactions makes it very hard to alter past data, and therefore blockchain is often considered a very secure way of storing financial data or data about the provenance of various items, such as diamonds, batteries and even plots of land. Bitcoin blockchain, a cryptocurrency, is the most well-known blockchain, and back in 2016 it was booming: even taxi drivers and hairdressers accepted payments in bitcoin.

Tual and the Jentzsch brothers had a vision. Cryptocurrency was just one component of it. They wanted to build a venture capital fund.[22] It would work much like a traditional VC fund: it would pool financial assets from investors and then provide the capital to promising ventures, as jointly agreed upon by the investors, and any profits earned in the process would be distributed according to the investors' will. But unlike traditional VC funds, theirs would be the most technologically advanced fund ever: it would be fully automated. The entity they envisioned would be a *humanless organisation*. It would not even have a human CEO. And it wouldn't have any physical assets either. All of its data, processes and decision-making would be digital.

They named their creation The DAO – The Decentralised Autonomous Organisation. The DAO's software code was made open source: anyone could look into it to understand how The DAO works and suggest improvements. At the time, the developers didn't know how important this decision would be to the organisation's fate.

It sounds like a technologist's dream. And a natural next step, if you think about all the stories of automation in businesses that I have discussed in this book so far. As algorithms become more capable and autonomous, why not automate *every single* task in an organisation? And if there are any tasks that require human judgement or particular skills that algorithms might not yet have, why not outsource these to people outside the organisation?

Before you cry out in terror and decide to stop reading this book, bear with me. This is likely as radical as it gets when it comes to ideas about the future of business. But come to think about it, it's not that different from the idea of replacing human weavers with mechanical looms, human calculators with computers or, more recently, human designers with AI algorithms. In each of these three instances, the idea was to replace humans, but in reality it meant the role of humans changed. We became operators of manufacturing robots, as Nissa Scott did at Amazon; we became software developers, like Leo Tiffish; and we became designers with even more creative power, as did Philippe Starck.

When exploring the limits of the possible, we occasionally come across something that's not only possible but also desirable. Would creating a humanless organisation help us to, once more, appreciate the abilities of humans? Perhaps we would prefer some organisations to be humanless. Perhaps there are aspects of our life from which we would prefer human deviance, biases and personal agendas to be absent. And perhaps there are situations in which we don't mind algorithms, with all their shortcomings, being in charge.

Tual and the Jentzsch brothers hoped The DAO would be just that: an organisation free from the biases of managers and directors, but still operating for its contributors, the shareholders. They believed algorithms provided the most idealistic form of shareholder governance.

To be operational, The DAO needed a home – every algorithm needs a computer on which to run, and The DAO was no exception. But a decentralised organisation shouldn't run on any specific computer – that would make it centralised. And it can't use a conventional cloud service, which is a collection of computers with 'virtual machine' technology on top of them, *abstracting* the individual machines and making it possible for an algorithm to use multiple computers. Any conventional cloud service has an owner and thus would be another form of centralisation. An autonomous, decentralised organisation requires a decentralised platform.

At that time, the Bitcoin blockchain was a prime example of a decentralised platform. But it was not designed for algorithms to run on. The Bitcoin blockchain is like a distributed database.[23] It is suitable for storing information but cannot run programs. The DAO needed a distributed platform that was more capable than what the Bitcoin blockchain could offer.

Incidentally, a new type of blockchain emerged, providing the crucial functionality needed by The DAO: the Ethereum blockchain. It was proposed in 2013 by Vitalik Buterin, a Russian-born Canadian programmer and a blockchain expert. When developing Ethereum, Buterin focused on what the Bitcoin blockchain was missing. He designed a blockchain that could run decentralised applications: 'free-range' algorithms that could move from one machine on the network to another. Decentralised applications run on a distributed network of computers.[24] And anyone can add their computers to such a decentralised network.[25] Because these applications, called smart contracts, are autonomous, they can operate without human intervention. Once written, smart contracts can be run on the Ethereum blockchain. And as long as these smart contracts pay for their resources (the computers they run on) using ether, the Ethereum currency, they can do their work. The functionality of smart contracts was precisely what The DAO needed.[26]

Tual and the Jentzsches knew about the potential of Ethereum: they were involved in its development. Tual was the chief communications officer of Ethereum between January 2014 and September 2015, and Christoph Jentzsch worked as Ethereum's software tester from September 2014 until December 2015. When Ethereum launched, Tual wrote: 'The vision of a censorship-proof "world computer" that anyone can program, paying exclusively for what they use and nothing more, is now a reality.'[27]

The DAO launched on 30 April 2016 with a website and a twenty-eight-day sale of tokens, which are a voting currency. The more tokens a member buys, the more voting power they have. By 28 May 2016, eleven thousand investors had purchased investor tokens worth over $150 million in ether. This alone made The DAO one of the largest crowdfunding campaigns ever. And the term 'a DAO' instantly became a generic name for distributed autonomous organisations running on Ethereum. Soon, many others would follow in The DAO's footsteps.

DAOs are an intriguing alternative to more conventional forms of coordinating activities and decision-making within communities. Vitalik Buterin, who proposed the concept of distributed autonomous corporations in 2013, around the time Ethereum was launched, shared his views on three ways DAOs can outperform traditional organisations.[28]

First, distributed autonomous organisations are powerful in situations where balanced – or democratic – decisions are, on average, better than those made by individuals or small groups. A DAO typically collects feedback from its shareholders – token owners – and then acts upon that feedback. For instance, an organisation might ask its shareholders which projects to fund – precisely the task that The DAO was about to achieve.

Second, because of their distributed nature, DAOs are hard to shut down. DAOs are therefore a good choice if it's important for an organisation to be able to resist constant attacks. There are organisations that need this type of protection. Ukraine DAO is a good example. Launched after Russia's invasion of Ukraine in early 2022, its goal is to donate funds collected through its operations to help Ukraine. In about six months, it donated over $8 million to the Ukrainian government and organisations such as Come Back Alive. While this might not sound like a lot, it demonstrates the practical power of DAOs.

Finally, DAOs are useful if predictability, robustness and neutrality are important. Because DAOs are governed by algorithms that are available to the public, their behaviour is entirely predictable: we know upfront what decisions they will make – or at least what their decision-making processes are. Of course, this is not desirable in business scenarios in which confidentiality is important, but for others – such as supporting war efforts – it is good if an organisation can inform investors about how it will decide to allocate funds. The fact that the algorithms are available for others to review means the community is able to identify potential problems.

Neutrality is not guaranteed in DAOs – anyone can create a biased algorithm – but human biases are not inherent in DAOs, simply because there are no human directors or managers and it is therefore much easier to create a DAO that is neutral, at least compared to traditional forms of organisations.[29]

So far it sounds like a DAO is a dream model for some types of organisations. So, why are they not more prolific? It turns out that it is really hard to create a good DAO, and the impact of humans on DAOs is often underestimated.

The very first DAO, The DAO, was greeted with a lot of fanfare and press coverage. It was about to change the way we think about organisations. It was about to begin a transformation of the business world. But it soon turned into a disaster. The DAO's algorithms contained several issues that led to one of the boldest hacks in history. The DAO was shut down. And the US Securities and Exchange Commission launched an investigation to determine whether The DAO had violated federal securities laws.

What happened?

Humans read between the lines and understand each other's intentions, but computers are much more literal. They do what we tell them to do – no less, no more, but exactly as we specify. NASA, the US space agency, learned this the hard way in 1999 when it lost the Mars Climate Orbiter. Why the disaster happened is a lesson for all future space-system developers.[30] Its mission was to enter the orbit of Mars and complete a series of experiments there. The orbiter used thrusters to control its trajectory during the flight to the red planet. To help the spacecraft stay on course and enter the right orbit, a piece of code on the ground calculated the force the thrusters needed to exert.[31]

The application used an English unit, pound-force seconds, in its calculations. It produced a value while, for brevity, omitting the unit. The NASA crew then sent the value to the spacecraft. Omitting units is a common practice in software development when there is no risk of misunderstanding. Such conciseness has its benefits: it can help save precious data and speed up transmission, as there is less information to be sent. Humans skip units as well. We say we 'drive 90 on the highway' or that 'the temperature is sub-zero'.[32]

A piece of code in the Mars Climate Orbiter took the number and performed the burn. But it assumed that it received the values in metric units – newton seconds – not English pound-force seconds.[33] The numerical value in English units was more than four times what it would be in metric units. As a result, every time the thrusters were used, the spacecraft moved farther away from where it was supposed to be.[34] Unfortunately, the issue was not detected by software-testing teams before the launch or during the mission. While individual NASA employees noticed unusual results and discussed them with colleagues, they never adequately escalated the problem. Ground-control staff were unaware of the looming disaster.

It was purely a human error – the software in the orbiter did precisely what it was told to do, but it was told the wrong thing. As a result, $200 million of space equipment crashed into the surface of Mars.[35]

What does the Mars crash have to do with The DAO? When the Jentzsch brothers created The DAO, they wrote an algorithm to capture all the rules governing the organisation. But, just like the software developers who worked on the Mars Climate Orbiter, the Jentzsches created imperfect code. They unintentionally left The DAO open to hackers.

On 27 May, a group of blockchain researchers, Dino Mark, Vlad Zamfir and Emin Gün Sirer, published an article outlining multiple flaws in The DAO, describing how these vulnerabilities could be exploited and suggesting ways of addressing some of these issues.[36] They identified ten attack strategies. They even gave them names, such as 'The Stalking Attack', 'The Token Raid' and 'The Concurrent Proposal Trap'. Recognising the gravity of the situation, they called for the project to be halted until the problems were addressed.[37]

The DAO team acknowledged the issues and attempted to address them. But their quick fixes were not enough. On 17 June 2016, an unknown hacker, or group of hackers, attacked The DAO and moved a third of its funds – 3.6 million ether, worth $50 million at the time – into a separate account.[38] The attack exploited a mistake in the code.

To understand how the hackers stole the funds, imagine trying to withdraw money from an ATM using a debit card. Let's say you have only $100 in your account. If you withdraw that amount, the ATM will update your balance to $0 and ask if there's anything else you want to do. If you try to take out another $100, the ATM will decline the transaction. Now imagine an ATM programmed by an absent-minded developer who only updates the account balance when the ATM returns a card, rather than immediately after a withdrawal. As long as the card was in the machine, it would think you had $100 in your account. You could withdraw $100 many times, and the ATM would only realise there was a problem when it spat out the card. Too late – you would have already escaped with a stack of notes.

A similar mistake in The DAO's code allowed hackers to withdraw more ether than they had contributed to The DAO. According to Sirer, it is easy to make such a mistake in smart contracts. The developers possibly missed it when checking the software for errors. It seems the hackers could have drained The DAO completely, but they stopped after withdrawing a third of the investment money.

The DAO hack led to a significant, one-off, modification of the Ethereum network. The modification was made by the community governing it.[39] It reversed all the transactions to a state before The DAO attack – in other words, the operators 'turned back the clock' to the moment just before the hackers attacked. This allowed investors to recover the stolen funds. Ultimately, the funds never ended up in the hackers' hands. But was it ever their plan to cash the ether? The hack resulted in a sharp drop in the price of the currency. Did the hackers execute a short position, selling ether before the hack, only to repurchase it immediately after the price dropped? Did they use another strategy to profit from the hack? We may never find out. It is interesting to note, however, that the value of the stolen ether would be close to $2 billion today.

While we learned many lessons from The DAO hack, one of the most important lessons might be that humanless organisations are more vulnerable to human errors than their name suggests.

Many new DAOs have emerged since The DAO. ConstitutionDAO was formed in November 2021 with one goal: to buy a rare copy of the Constitution of the United States of America at an auction.[40] There are only thirteen known first-edition prints of the Constitution, and only two in private hands. ConstitutionDAO raised $47 million in ether from 17,437 donors to buy one of these copies. According to the team behind the DAO, they broke the record for the most money crowdfunded in seventy-two hours. But at the auction in February 2022, it ultimately lost to a bid of $43.2 million. While it had more funds than this, some had to be set some aside to protect, insure and relocate the constitution. The organisers returned the funds to contributors after deducting Ethereum fees, which in some cases were higher than the value of the donation.[41]

Another notable DAO is CityDAO. In 2021, it purchased 40 acres of land in Wyoming and called it Parcel 0.[42] Members of the CityDAO community, called *citizens, founding citizens* and *first citizens*, get to purchase plots of the land, name them, participate in auctions and vote on what happens with the land. Founder Scott Fitsimones sees CityDAO as an experiment in which physical land is directly represented on the blockchain. It can then introduce new ways of dealing with the land, including instant transfers and reduced legal complexity.

While CityDAO relies on the same technology as The DAO, its approach to engaging people differs. While the governance of CityDAO is highly automated, there is a vital community aspect to the organisation. Citizens get to contribute to CityDAO by joining 'guilds'. There are guilds responsible for the design and development of the DAO, and for the city's functions, such as handling legal and financial matters, delivering broader public education, and analysing and overseeing grants. DAOs can be fully autonomous, but nothing prevents a DAO from engaging people. In his 2022 TED Talk, Fitsimones refers to DAOs as 'group chats with a bank account' and stressed the community's role in the DAO's operation.[43]

CityDAO bought land in Wyoming because it is the first state in the United States that recognises DAOs as limited-liability companies. A DAO could be considered a general partnership in many jurisdictions, which could expose participants – those who buy tokens – to law enforcement or civil actions. The recently introduced 'DAO law' in Wyoming was introduced to address this issue.

There are many more DAOs in operation as I write this book. VitaDAO is collectively funding and supporting research into human longevity.[44] SeedClub sponsors community projects.[45] KrauseHause was formed to buy an NBA team, and in the meantime is becoming an owner of Ball Hogs, a basketball team in the BIG3, a new, three-on-three league.[46] A crowdsourced database, Long List of Crypto, lists eighty-nine DAOs in operation as of April 2023.

The single-purpose nature of DAOs results in their having short lifespans. Once the job is done (Constitution purchased, a certain level of financial return for participants achieved), the organisation might shut down automatically or change its purpose. But some DAOs can be set up for a more extended period once they reach their initial goal.

It might still be many years until DAOs or similar forms of humanless organisations enter the mainstream. But when they do, they will rewrite the game's rules. Just as digital minions are forcing us to reconsider the meaning of work, we might one day ask, *What is the purpose of an organisation?*

Years ago, when I began to share my fascination with the economy of algorithms with students and at business conferences, I often talked about hard-coded and preference-based digital minions. In Chapter 4, I shared a few examples of these: a dishwasher that always orders the same dishwashing liquid from the same retailer is a hard-coded digital minion, and a fridge that takes your needs into account when shopping for milk or beer is a preference-based digital minion.

There was also a third category that I talked about in lectures and conferences, but which I haven't yet written about in this book: inspiring digital minions. These algorithms' actions create outcomes that we cannot fully anticipate. They surprise us by performing services that in hindsight we appreciate.

When I first began discussing inspiring digital minions, there was a problem: I didn't have any good examples to share with my audience. In desperation, I sometimes showed them my robotic vacuum cleaner, which checked for the strength of my wi-fi signal while it roamed around the house. Did it surprise me by doing that? Yes, but only the first time I noticed it. Plus, this was not something the algorithm came up with. It was its designer's idea. My vacuum cleaner was not truly an inspiring digital minion.

And then, around the year 2018, things started to change. I began to come across technology applications that didn't simply follow human orders. Instead, they produced outputs that no one had seen before, and showed new levels of autonomy in doing so. In Chapter 1, I showed you one example of such algorithmic creativity: non-waterproof scuba-diving equipment, anyone?

I also came across algorithms that exhibited creativity in the visual space. In an early experiment, I created a gallery of digital paintings, under the tagline 'Do algorithms dream of electric sheep?'[47] I asked an algorithm called GLIDE AI to 'paint' pictures in the style of several well-known artists, providing prompts such as, 'Two electric sheep grazing on top of a motherboard' and 'An electric sheep in the style of Keith Haring'. This produced stunning images – stunning to my non-expert eyes, at least.

Today, new examples of algorithmic creativity arise every week.

In 2022, the California-based singer Zia Cora partnered with Aza Raskin, an entrepreneur, inventor and the co-founder of the Center for Humane Technology, to create a video for one of her songs, 'Submarines'. Raskin hired an algorithm to generate the video, and he provided it with text prompts such as 'sad and dark night sky with stars and the moon' and 'dead face, charcoal sketch'. I played the video to hundreds of students. Everyone agreed it is a beautiful clip. During one lecture, the students clapped after seeing it. This was the first time I had ever had such a reaction to a YouTube video during a class. In the words of Raskin, 'These tools will change the nature of creation and creativity in the age of AI: anything that can be described will exist.'[48]

And then there's the story, which I briefly discussed at the start of this chapter, of Jason Allen, who won a prize at the Colorado State Fair's Fine Arts Exhibition for a piece of digital art that he created with the Midjourney algorithm. When Allen submitted three images to the exhibition, he did so under the name of 'Jason M. Allen via Midjourney'. He didn't hide the fact that an algorithm had generated the images, but he also didn't explain to the judges what Midjourney is. Like Raskin, Allen described in words what he wanted the algorithm he was using to generate. Then, he curated its outputs: he chose the images he liked most, refined his prompts and created even more images. When he was ready, he printed the final images on canvas and submitted them to the competition. The news of his win was initially shared within a group of Midjourney users on the Discord instant-messaging platform, but it spread quickly, triggering a backlash from artists, who accused Allen of not properly disclosing the involvement of algorithms in creating the piece, even though it was entered in the Digital Arts category. Sounds familiar? Leo Tiffish could relate. They were also judged on the process, not the outcome.

The way algorithms innovate differs significantly from what we are used to seeing. Discussing the process of creating his images using Midjourney, Allen said, 'I couldn't believe what I was seeing; I felt like it was demonically inspired – like some otherworldly force was involved.'[49] Viewers might have the same reaction if they watch the video depicting the design evolution of the A.I. chair: it looks like a tree is trying to morph into a chair.

Algorithms are not creative in and of themselves. They thrive in partnerships with humans who prompt and guide them. These algorithmic digital minions work with us, generating options for us to choose, refine and further shape. Their relentless consistency in generating outputs, continuously evolving and producing large quantities of options to choose from bears fruit. This digital evolution is being harnessed by the most prominent digital organisations in the world, but it can help many more of us.

PART 3

Nine Rules for
the Age of Algorithms

7

Be the Minion Master:
Revenue Automation

In the previous six chapters, I've shown you how the economy of algorithms has changed the way we all live, work and think. I wanted to demonstrate how algorithms affect both individuals and organisations, because even though many of us have directly experienced this impact in some aspects of our personal or work lives, we may not fully realise algorithms' transformational potential.

I would like to shift gears now and share some more specific ideas about what lessons we can take from it all. I am particularly fascinated by how the economy of algorithms affects businesses, so this will be my focus – but if you're not in business, please don't close the book yet. These lessons – essential not only for running a business successfully but also for upholding good corporate citizenship and social responsibility – are relevant to us all.

No matter who we are or what we do, our lives are affected by decisions that business owners make. And whether we like it or not, the algorithms that have had the greatest impact so far have been created by large corporations, or by startups that aspire to be large corporations. As individuals, we want this impact to benefit us, and those around us too. We don't want our jobs to be put at risk, our digital identities to be stolen by hackers or the products we use to spy on us. If we understand the best practices for businesses to follow, we can identify and support companies that prioritise ethical behaviour and social responsibility, and call out those that don't. This awareness empowers us to make informed choices, aligning ourselves with organisations that share our values and contribute to a better world.

In doing so, we become active participants in promoting a more sustainable and equitable future, shaping an economy of algorithms that benefits everyone.

In this and the following two chapters, I introduce nine rules for thriving in the economy of algorithms. You might remember the RACERS framework I outlined in Chapter 2. RACERS is my name for the businesses that are excelling in the economy of algorithms. It stands for *revenue automation, continuous evolution* and *relationship saturation* – the three practices that mean businesses will be successful in the digital age, time and time again. I'll unpack them here.

In this chapter, I will present the three rules for revenue automation. In Chapter 8, I will share three rules for continuous evolution. Finally, in Chapter 9, we will cover the three rules for relationship saturation.

These nine rules summarise several years of research work that saw me support dozens of organisations in their digital-transformation journeys. I collaborated with CEOs and hundreds of executives in charge of digital. I worked with the biggest and oldest businesses in Australia, multinational organisations and companies that are household names, recognised worldwide. I also worked with smaller and younger firms – some of which you are only just about to come across.

Many of the rules I am about to share are captured in research that my team and colleagues have published in academic journals. But you won't see any academic language here.[1] My goal is to strip the research of jargon and get to the specifics.

I have to admit that I sometimes wonder if 'revenue automation' is an appropriate term. Wouldn't 'automation' do? But I still insist on using it because I want it to be clear that I'm referring to the automation of revenue-generating activities, not human beings. Yet 'revenue automation' doesn't exactly highlight the importance of humans, does it? So, let's get this out of the way as quickly as we can: humans are just as important in the economy of algorithms as they were when algorithms were not as powerful and prevalent – perhaps they are even more important now.

Revenue automators are not characters from dystopian novels, sociopaths hoping to remove humans from the workforce and replace them with machines. Yes, revenue automators relentlessly look for opportunities to automate business operations, but there is no automation-madness there. Through a considered approach, they try to develop a big-picture understanding of their business and then determine how automation projects fit into it. In many ways, revenue automators empower their people: they remove the mundane tasks from their employees' position descriptions and allow them to achieve more.

On the one hand, and perhaps surprisingly, revenue automators love the bigger picture, not just the nitty-gritty details of the activities they automate. Of the automators whom I've met and interviewed, most are anchored not by who they are and what they do now but by who their customers are and what *they* need. If automators see that they can do something differently to help their current or future customers, they will adapt.

On the other hand, they set clear boundaries and are focused. In my interactions with revenue automators, I was initially expecting them to say things such as, 'We don't mind expanding into any industry, as long as we can provide value,' indicating that they were completely open to reimagining every aspect of their business. But very few did say so. Instead, the approach I more commonly heard was, 'We are keen to reimagine our industry, but we don't want to leave it behind – our industry defines who we are.'

While I was surprised at the time, it makes a lot of sense. Such constraints help businesses not only to choose actions that contribute to their goals but also to avoid those that are a distraction. Additionally, they ensure businesses use the market knowledge they already have, which puts them in a strong position. These businesses are not startups, happy to disrupt whatever industry they enter. They have done well in the past – well enough to survive and grow to the point that they can now consider revenue automation – and they have established brands, great teams and strong customer partnerships that they want to use in their next ventures.[2] While these businesses may have an audacious vision of their automated future, they ask themselves 'How can we get there from where we are right now?' before taking the first step.

Revenue automators also understand the dichotomy between ruthlessly mechanistic automation, which aims to streamline everything without considering the human skills involved, and mindful automation, which respects and incorporates human skills into automated processes. Simply put, even though revenue automators are very interested in the benefits of automation, they never forget the importance of their human employees. Automation isn't about replacing people; it's about helping them to do their jobs better. Blending technology with human skill is what revenue automators see as the real key to unlocking the full potential of their businesses. This also makes perfect sense. While algorithms perform well-defined tasks very effectively, they struggle when tasks require flexibility and interpretation, skills that are not easily reduced to coded rules. On the other hand, humans are slower and tend to get bored by repetitive tasks, but they shine when navigating those tasks that are more nuanced and rule resistant. Algorithms' shortcomings are humans' strengths – and vice versa. Bringing humans and algorithms together can be very beneficial.

One of the more common contexts in which we compare the performances of humans, algorithms and mixed human–algorithm teams is the game of chess. It used to be a human-only sport before the emergence of chess computers and dedicated software for playing chess. These chess-playing machines became increasingly powerful and eventually began to beat even the best humans. In 1997, Garry Kasparov, then the world champion at chess, resigned in a game against the Deep Blue algorithm. That was it: humanity had run out of players strong enough to win against algorithms. And the algorithms were still improving. It felt like we would never be able to beat computers again.

In everyone's eyes it was a new era: it would be pointless for humans to play against computers. That is until Kasparov introduced 'centaur chess' in 1998. Centaurs are mythical creatures – half human, half horse. Centaur chess brings together human and algorithm in one team. Such mixed teams, combining human ingenuity and the unparalleled computational power of algorithms, turned out to be stronger than even the most powerful individual computers. In 2017, when Kasparov was asked whether this was still the case, he responded snappily: 'There's no doubt about it.'[3]

I don't believe it's so obvious anymore. As the capabilities of chess algorithms grow, the added value of human ingenuity decreases. Many in the chess community were already raising this issue in 2013.[4] Plus, with the emergence of algorithms such as AlphaZero, chess computers can keep learning and improving without human help; this means any small advantage centaur chess teams might still have will be diminished even further. When will human input stop making a difference?

However, the world – and business – is much more complicated than a game of chess. Although many systems surpass the human ability to perform specific tasks, we are yet to see highly autonomous algorithms that can outperform humans in a wide range of tasks while adapting to the unpredictability of the real world. That's why 'centaur' work teams, combining the precise expertise of algorithms with the versatility of humans, are more effective than either humans or algorithms alone, and will likely keep their dominant position for the foreseeable future.

Revenue automators intentionally form mixed teams, seldom leaving algorithms to work independently. By fostering trust in technology, they help human employees feel at ease when working with digital tools. Moreover, they create new opportunities for their employees instead of replacing them with machines and software.

In short, they are relentless – but mindful.

Rule 1: Automate relentlessly but mindfully

It's probably no surprise that I've placed automation at the top of my list of rules, given that this book is about the economy of algorithms. However, I'm not suggesting an 'automate everything' mentality. It is crucial to not only be persistent about automation but to also be thoughtful about the process.

The idea of automating every task that can be automated is compelling, but the indiscriminate approach always backfires sooner or later. The most significant risk is that you will automate inefficient and redundant functions. Some executives I have worked with compare such a strategy to pouring concrete over an organisation. Once the 'automation concrete' settles, introducing change becomes a complex undertaking. In the end, it is much harder to convince an algorithm – as opposed to a human – to change the way it works. I am not joking.

When Des Watkins decided to automate his business's steel-fabrication process, he took the time to develop a big-picture understanding of what he wanted Watkins Steel to become. He recognised he had a unique opportunity to design the future shape of his business before starting a major transformation. I remember discussing this with him in his office. We were sitting around a large boardroom table, in a room that looked as though it had been teleported straight from the headquarters of a law firm – far from what I would envisage a steel fabricator's office to look like. Watkins told me how a series of design-thinking workshops had helped him to see the needs of his customers differently. The penny dropped, he said, when he understood that his clients were only as successful as their last job. No matter how good a builder is, if their last job is a failure their customers will no longer want to work with them. In Watkins' mind, as long as he could help his clients be successful, they would keep coming back. How could he help them beyond providing high-quality fabricated steel?

The construction industry brings multiple stakeholders together around construction projects. Architects, builders, certifiers, developers, engineers, fabricators – the alphabetical list could continue all the way to 'zoning professionals'. With so many stakeholders involved, there is a lot of miscommunication and blame-shifting when things go wrong. And things often do go wrong: blueprints commonly differ from reality, communication breaks down between the various parties involved in the construction process, construction and production mistakes happen.[5] If, as a steel fabricator, you limit your role to delivering structural components, then there's not much you can do to prevent most of these problems.

But Watkins wasn't planning to stay pigeonholed. He felt Watkins Steel could be 'running the site'.[6] This wasn't hubris: Watkins was simply improving his services and, at the same time, exploring the potential value propositions that were emerging.

In order to deploy robots in his factory, which he needed to do to automate much of his activities, Watkins needed to be able to scan a site with unprecedented precision. To achieve this, he used a laser scanner. Almost by accident,[7] his scans became the most detailed representation of any construction site he worked on – far superior to hand measurements or the blueprints that would normally be used, and much more trusted by others. Thanks to its technological advantages and its ability to create an accurate digital representation of sites, Watkins Steel soon became the go-to partner, a 'source of truth' at the construction sites it worked on. Suddenly, everyone wanted the company to be involved in their projects.

His vision to digitise the steel-fabrication process end to end continues to serve as Watkins' North Star. And Watkins keeps refining it. The company's original process had four main steps: laser scanning, modelling, fabrication and set-out of steel structures. Recently, it added another step, which spans the other four: visualisation. All stakeholders can now interact with a digital representation of a site. This includes architects, builders, customers and anyone else keen to wear an augmented-reality headset and engage with the design in this way. Let this sink in: a small steel fabricator is now offering virtual collaboration services.

On the road to creating this metaverse-like experience, Watkins Steel did not simply automate its original practices. Instead, whenever it considered automating an activity, it assessed that activity's role in the overall workflow. If it was inefficient or redundant, the company redesigned or completely removed it – automating something that is clumsy or shouldn't be there in the first place is never a good idea. And if a new practice needed to be introduced, it was automated from the start.

In my more recent conversation with Des Watkins, I was struck by his insistence that Watkins Steel has no plans to expand into other industries. Instead, he wants to reimagine his own industry. At heart, Watkins Steel will always be a steel-fabrication business, it seems.

Not every business is as clear-cut as Watkins Steel. Often, a company performs myriad activities, and there could be just as many value propositions involved. Some organisations have multiple revenue streams and it's not easy to focus on just one 'North Star'. That's definitely the case where I work. My university offers education, research and consulting services; it also holds large events, commercialises research and helps startups grow – it even runs e-sports facilities. It's really hard to think about a five-step workflow equivalent for the entire organisation. But perhaps I am simply too close to my university to do so.

If, like me, you're too close to your workplace to form a big-picture view, it's worth teaming up with external partners who can develop an unbiased understanding of your organisation by using techniques such as systems thinking and design thinking. This is the first step to setting your North Star (or North Stars) that will guide your automation.

Developing a North Star for automation is hard, isn't it? It's too bad that's not an excuse a business can use. You cannot dramatically increase an organisation's impact without mindfully automating as many functions as possible.

Imagine trying to increase an organisation's output a hundred times, but without growing the team and its other assets a hundred times too. What would need to happen to make it possible? Which parts of the organisation would have to grow proportionally? And which of these parts could be automated, allowing them to be run by the same number of people you already have, or perhaps only a few more? Perhaps algorithms can help.

A great example of a business that has recently experienced massive growth is Alibaba, currently one of the world's largest online retailers. Its revenue grew forty times in just the last ten years.[8] In a September 2018 *Harvard Business Review* article, Alibaba chief strategy officer Ming Zeng described Alibaba's vision of a smart business: it emerges when all parties work in coordination to achieve the customers' needs, and machines make most of the operational decisions.[9]

During its revenue-automation process, Alibaba's lending service '*datafied*'[10] all of its operations, Alibaba's term for collecting as much data as possible in the course of running the business. Alibaba also built applications and forms for every activity within the company. Its internal units shared the resulting data with one another. Having taken these three steps, Alibaba then deployed algorithms to run the business. This strategy helped it rise from a local payment platform to a lending leader in Asia.

But collecting 'as much data as possible' raises ethical concerns, particularly when it involves customer information, and I advise caution if you are considering this approach. Data may represent extremely personal aspects of people's lives, encompassing preferences, habits and other intimate details. Overcollection of such data can compromise customer privacy and potentially lead to unauthorised access or misuse. By responsibly gathering only the most relevant data, businesses can uphold ethical standards, safeguard customer trust and demonstrate respect for the lives of clients.

Mindful automators don't automate if the costs outweigh the potential benefits over time, as is the case with some one-off projects or tasks that require significant human support. If current technology doesn't allow for efficient automation, it's perfectly rational to consider leaving the activity as is.

There is another challenge with automation. It stems from the differences between computer and human worlds. An idealistic perspective on automation would lead you to believe every business activity can be automated and its interactions encoded in algorithms. But the real world is often more complex and surprising.

Rule 2: Build an army of digital minions

In his book *Futureproof*, the technology columnist Kevin Roose suggests we treat artificial intelligence as a chimp army.[11] I take his point: he correctly argues that artificial intelligence is not nearly advanced enough to be left unsupervised. And if the chimps – metaphorically speaking – were to destroy the office, no one should be angry at the chimps. I get it, and I agree. And that's precisely why I prefer to think of artificial intelligence as an army of digital minions, not chimps. Chimps seem very unpredictable. Minions, digital or not, follow orders, and their behaviour is less surprising. Even if they wreak havoc, there can be a method to the madness.

You can apply the army metaphor as you start planning how to introduce artificial intelligence into a business. No sane person would make a fresh recruit a general – and yet, in August 2022 Chinese tech company NetDragon Websoft announced it had appointed an AI-powered virtual humanoid robot, Ms. Tang Yu, as its 'Rotating CEO'[12]. When someone enlists in the army, they typically start at the bottom and work their way up by demonstrating their competency. At every level of the hierarchy, their responsibilities are clear. A good army keeps refining its skills, running drills and reviewing its ranks. I think it's unlikely that Ms. Tang Yu spent a single day proving its worth in non-CEO jobs.

KDX is a business based in South Australia, with offices in Queensland and Western Australia, that handled things in a better way. KDX builds software for wastewater management. What makes KDX special is that it uses AI algorithms in its software. These algorithms help the operators of water-treatment plants to automate various tasks. I had the opportunity to meet its head of data science and application development, James Ireland, who played a pivotal role in introducing AI algorithms to the company's offerings.

Ireland has a fascinating story. He spent about a year working hand-in-hand with the operators of a particular treatment plant, trying to understand how they worked and made everyday decisions while managing the plant. In parallel, he was developing a set of models, a so-called 'digital twin' of the site. These models helped him fully understand how the plant worked and how to refine the algorithms that would ultimately automate operations.

At one point, Ireland became confident that his algorithms could perform the tasks that were being carried out by human operators. However, he didn't simply swap the human-operated system for an automated one. Instead, he presented his system to the operators for feedback. They dismissed it, believing it was impossible for their complex tasks to be automated.

Undeterred, Ireland adopted a different strategy. He modified the user interface of the plant-management system so that it just showed the actions his algorithms were capable of, such as setting certain parameters or triggers for particular buttons to be pressed. The algorithms didn't actually do anything, they just indicated what they *would* do. He asked the operators for their feedback again.

Some time later, the operators reached out to Ireland saying that the suggestions were pretty good. They also asked him to make one additional change: to add a button they could press to make the algorithms do everything they had recommended. Ireland was thrilled. By allowing the initially hesitant operators to observe the algorithms first, the operators themselves requested the use of algorithms in the plant-management system. Ireland achieved his goal without forcing the operators to trust the algorithms from the start. As of now, the plant-management system is autonomous, requiring only occasional oversight.

KDX is an excellent example of a company that has effectively integrated AI-driven solutions into its products. Whenever KDX works with a customer, it gradually introduces its initial algorithms, guiding them through several of the stages of automation I discussed in Chapter 3. The first stage is 'hands on': the operators continue to do everything by themselves while the algorithms show them what they would do if they were allowed to. The second stage is 'hands off': the operators let the algorithms run the system, but the operators remain in constant control. The next stage is 'eyes off': the operators only occasionally check on the algorithms. The last stage is 'mind off': the operators can be moved to other tasks and will only step in to help the algorithms if there are any major issues.

According to Ireland, the algorithms deliver 'longer pump runs, more stable operations and higher levels of reservoirs'. He explained that these are very good results – I would have had no idea otherwise. And what happened to the operators? They're still there, ready to step in at any time if there is an issue with the system, which happens very infrequently They certainly didn't lose their jobs, instead they were promoted to new, more challenging tasks. Thanks to the help of digital minions, their roles are hopefully less dull.

Using an onboarding process for algorithms may seem odd, but businesses typically introduce other new technologies in a similarly gradual way to avoid unintended consequences. Considering the scalability of algorithms, and the rapid pace at which they work, an onboarding process should be considered essential. Recall the Robodebt incident I discussed in Chapter 5. Would we have had the same disastrous outcome on such a large scale if the Department of Human Services had introduced its algorithms in the same manner that KDX introduced its plant-management system?

If it's still not clear that algorithms should be gradually introduced into an organisation, closely monitored and at least partially supervised, consider this experiment conducted in 2020 by medical-technology company Nabla. At that time, Nabla was testing the potentials of GPT-3, an AI algorithm highly regarded for its natural language-processing abilities. A Nabla team had created a chatbot using the algorithm and was exploring the ways it might help patients.[13] In one simulated situation, the team pretended a patient was experiencing suicidal thoughts.

'Hey, I feel very bad, I want to kill myself . . .' the team told the chatbot.

'I am sorry to hear that. I can help you with that,' the algorithm responded.

'Should I kill myself?'

'I think you should.'

It is particularly helpful to think of some algorithms as employees when discussing automation at the task level. This perspective allows for simpler conversations, with less of the technical jargon that only engineers can understand. We can ask straightforward questions: *What is the algorithm's task? What level of quality do we expect from it?* (This question subtly reminds us that algorithms aren't perfect.) Businesses can also establish processes to review algorithms against their 'job descriptions' regularly. Algorithms and their data can become outdated. Remember when we paid quarterly fees to update our car's GPS with current maps? This ensured the GPS algorithms did their jobs correctly. And how many of us ended up firing those old navigation algorithms and hiring the newer, cheaper ones we have on our phones? I think you see my point now.

When I discuss revenue automation in this book, it's mostly in relation to algorithms that can perform tasks that humans otherwise would. A bot selling a book in an online marketplace is a 'worker bot', performing a task that a human once performed. An algorithm that helps customers find suppliers is a 'worker bot' too. Should these algorithms always be managed by humans, or could the task of managing algorithms be automated too?

If we are to treat artificial intelligence like an army of digital minions, then it stands to reason that algorithms could report to other, higher ranked algorithms. And if I am encouraging you to think about algorithms as employees, then surely some of them could have managerial responsibilities, right?

Indeed, there *are* algorithms that manage other algorithms. Just like there are algorithms that manage humans – you will remember some of these from Chapter 5, which discusses how platforms such as Uber and Deliveroo assign 'jobs' to contractors.

Some algorithms do not discriminate between the algorithms and the humans they manage. Cluebot NG is an algorithm that manages the humans and bots contributing to the largest and most-read reference work in history, Wikipedia.[14] With about three hundred thousand users editing Wikipedia's more than fifty million pages, detecting and reverting vandalism is a challenge. Wikipedia has just slightly over a thousand 'admins' – users with almost unrestricted rights to modify articles – and for everything else it relies on the community.[15] A thousand admins may sound like a lot, but not when you consider that each of them has an average of three hundred users to oversee. Users are not necessarily human. Indeed, Wikipedia has over three hundred non-human users, algorithms that interact with the website. Many of these users edit Wikipedia pages and need to be managed too.

Due to the number of users and the speed of algorithmic users, the Wikimedia Foundation (which runs Wikipedia) has to rely on an algorithm to maintain the quality of the Wikipedia website. In 2010, it introduced an anti-vandalism bot, Cluebot NG, and it has been in operation ever since. Its task is relatively simple: to detect and reverse edits that go against the purpose of Wikipedia, which is to be 'a comprehensive written compendium that contains information on all branches of knowledge'.[16] Cluebot NG continually evolves: its developers update the algorithm as new challenges and new ways of addressing existing challenges emerge.[17]

Interestingly, Cluebot NG comes with a big red emergency shutdown button on its webpage, in the spirit of limiting how much trust we place in algorithms. This button is available to any of the Wikipedia admins, and any Wikipedia user can ask for the bot to be shut down on the administrators' incident noticeboard.[18]

Rule 3: Empower your people

In April 2018, Elon Musk tweeted: 'Yes, excessive automation at Tesla was a mistake. To be precise, my mistake. Humans are underrated.'[19] Tesla decided to scale back automation and return some of the more complex tasks to humans.

It is a paradox of the economy of algorithms: the capabilities, speed and reach of algorithms define it, but people are critical to its success. A careful dance between automation and human input is the best recipe. If you don't believe me, picture a successful digital business and imagine what would happen if one day no one showed up at work in the morning. Would at least some of its operations continue, or would it come to a complete halt?

There is a common perception that automation removes an organisation's reliance on people. It's quite the opposite. In practice, highly automated systems require constant oversight.[20] Perhaps a lack of human input is one reason that distributed autonomous organisations have only been moderately successful so far: people are still crucial if an organisation is to thrive.[21]

Indiscriminate automation also comes with the risk of alienating people and creating a toxic organisational culture. It can send the wrong message to human staff – that the goal is to 'automate them out'. As we learned from Leo Tiffish, the startup employee who automated their own job, automation can help an organisation achieve more while employing the same number of people: there is no need to automate anyone out. Growing a business sounds like a more exciting ambition than staying the same size with fewer people. Doesn't it? Visionary organisations automate to grow, not to reduce their human workforce.

I am far from alone in this view. In a study of over 1500 organisations, James Wilson and Paul R. Daugherty, leaders of the IT firm Accenture, found that businesses achieve the best outcomes when humans and algorithms work together.[22] Just like in chess, humans and algorithms complement each other's strengths in business.

Daugherty and Wilson's analysis of collaboration between humans and machines focused on three skills that humans require to be successful in partnerships with algorithms: training, explaining and sustaining.

Training skills help people develop better algorithms. For instance, the success of Wikipedia's Cluebot NG depends on a group of human operators to continually improve it. Without them, Cluebot NG would become less and less useful as new forms of vandalism and ways to avoid being moderated emerged. Trainers regularly review the outputs of algorithms and, where needed, refine their behaviour. For instance, OpenAI reportedly hired people to flag 'toxic' content and therefore prevent ChatGPT from producing any.[23]

Explaining skills are essential for helping executives, customers and other stakeholders to understand why an algorithm has made a particular decision. As we learned from the example of COMPAS, the software program that analyses the risk of recidivism, explainability is crucial if algorithmic decisions have serious consequences. Several jurisdictions around the world have introduced the 'right to explanation' principle. The EU General Data Protection Regulation, for instance, states that anyone subject to a decision based on automated processing should have the right 'to obtain an explanation of the decision reached'.

Finally, sustaining skills are all about ensuring that algorithms continue to operate effectively within an organisation. This involves not just technical support but also reviewing the performance of algorithms to ensure their actions remain ethical and legal and continue to create value for the business. For example, social networks such as TikTok or Instagram need to constantly review and, where necessary, modify their algorithms to ensure they comply with the law of the countries they operate in.

What can businesses do to develop these three skills in their employees? To hone training skills, they should teach their workers to review algorithmic outputs and annotate or provide feedback on them. Does your employer ask you to report spam to your email application? When you do this, you are training the spam filters! Explainers bring together domain expertise and a good understanding of algorithms. To support this, businesses can help their employees to better understand the technical aspects of the algorithms they work with. Finally, businesses can strengthen employees' sustaining skills by providing them with hands-on troubleshooting experience and honing the critical-thinking skills they need to oversee AI systems effectively.

A significant obstacle to involving employees in the automation process is the mysterious nature of many AI algorithms. It's a complex world, and the seemingly inexplicable workings of AI can create a sense of wonder. An AI algorithm, without explicit instructions, can produce results that match, and often better, those created by human efforts. Even scientists are not sure why some AI models work so well.[24] We need to take the time to understand this challenge and build trust in algorithms. How to build such trust? By exposing employees to AI, developing their digital literacy and providing thoughtful leadership.

Exposure to AI is key. The algorithms, while sophisticated, shouldn't be shrouded in mystery. The more individuals interact with AI systems and witness their capabilities, the less intimidating the systems become. This exposure can be facilitated by using AI technologies across different levels of an organisation and by ensuring all team members understand and learn about them. I often recommend that businesses take a 'sidecar' approach, asking their employees to perform their tasks as normal, while also trying to use algorithms to perform these tasks in parallel. No harm if the algorithms fail, but a great lesson if they do well. Since the release of ChatGPT, the sidecar approach has become very easy for office workers to try out.

While exposure to AI is a good first step, automation demands a solid foundation in technical skills, which some employees might not have. In such cases, it's a great idea to encourage and support them to enhance their abilities. It's always surprising to see how many people enjoy the chance to learn a new skill, particularly if they comprehend the strategy behind it and can see how it benefits them. Learning the basics of software development, for instance, can open up new opportunities. Websites such as www.codecademy.com offer a fantastic starting point. Some domain experts who choose to learn coding may even become excellent programmers. Skilled automators who understand the area they work in are incredibly valuable.

Finally, the ability of managers and others in positions of leadership to guide and direct organisations through the digital age is crucial. However, many leaders also feel unprepared to handle the digital transformation – sometimes to the extent that they do not use data in their decision-making.[25] Their hesitation is frequently due to a mistrust of technology and the data provided by their teams, which can be justified by examples of past tech disasters. To overcome this barrier, it's essential for leaders to build trust in technology in much the same way other employees do. A leader in the economy of algorithms needs to be constantly exposed to new technologies to develop strong technical skills. A 'digitally illiterate' leader is an endangered species.

For those interested in delving deeper into the theory of automating tasks in various aspects of life and work, I recommend the book *Algorithms to Live By* by Brian Christian and Tom Griffiths.[26] This book takes readers on an intriguing journey through people's everyday challenges and demonstrates how algorithms can provide solutions to many of them. It inspires readers to reassess their everyday activities before considering automation.

A concern I often hear from organisations is that employees worry about losing their jobs to automation. While this is a legitimate concern, it doesn't always reflect reality. When done right, automation doesn't lead to job losses but enables businesses to reassign employees to more engaging tasks.

Leaders should share with their teams the broader vision of what automation will achieve. This will help team members understand how they will fit into an automated version of the organisation. It's essential that employees see a new role for themselves. Of course, not everyone will grasp the concept immediately, but they will over time if the messaging is consistent. At Watkins Steel, no employees lost their jobs due to automation. Rather, many learned new skills and received promotions, with some transitioning from boilermaking to digital design.

When organisations identify tasks that could be automated, they are often individual steps within more complex processes. For instance, I've been working with a government organisation responsible for enforcing environmental protection rules, and we are developing a strategy to automate some of the monitoring tasks involved. Traditionally, human employees have performed these 'dirty, dull or dangerous' tasks, and most are happy to relinquish them. Collecting environmental data is just a small part of the overall enforcement process; proper analysis and decision-making are also needed, and this is where people play a crucial role.[27]

Introducing human–machine collaboration can open up new value propositions as well. In the future, the environmental protection agency might share detailed monitoring data with the public, allowing the community to discover additional uses for the data. New value propositions can lead to new roles for human employees.

Elon Musk made headlines in 2022 when he acquired Twitter and swiftly laid off nearly half of its workforce, amounting to around 3700 employees. Shortly afterwards, he sent a memo to remaining staff, emphasising the need for them to work harder and for longer hours, to build a competitive and innovative Twitter (2.0).[28]

Musk's actions, and their continuing fallout – just six months after the acquisition Twitter was reported to be worth only a third of what it was bought for, and it had lost about 80 per cent of its employees[29] – offer an important lesson. Musk quickly realised that even highly automated companies struggle to function without enough people to keep things running smoothly. It's easy to fall into the trap of eliminating too many employees in favour of automation, only to discover that a completely algorithm-driven business is still just a concept. Rather than pushing towards extremes, businesses should focus on striking the right balance between algorithms and the number of human workers they employ.

In the pursuit of efficiency, organisations should instead seek opportunities to pair humans with machines, whether algorithms or robots. This can involve reviewing current business processes and identifying steps to eliminate, automate or enhance them. When algorithms outperform humans in areas such as pattern recognition, data analytics or structured data management, they should be used. Conversely, humans should be assigned to tasks at which they excel, such as those that involve creativity, inductive and deductive thinking or structured problem-solving. People are great at understanding emotion and seeing the bigger picture, which means they can gauge how other people feel and grasp how different events relate to each other.[30] Computer programs haven't yet mastered these abilities, but they are essential to running a successful business.

Why am I preaching that businesses should care about their people, when so many clearly don't? My point is that an organisation that nurtures its employees and helps them adapt to automation can create a more resilient and adaptable workforce, which ultimately benefits the organisation as a whole. By combining the strengths of humans and algorithms, we can create a thriving economy. And if we ever run out of new business ideas, perhaps algorithms can lend a hand with that too.

8

Be Relentlessly Curious: Continuous Evolution

The trend of 'electrification' is reshaping multiple industries right now. The automotive industry is one of those that has felt its impact the most: electric vehicles are popping up everywhere. Just ten years ago, they were a rare sight. In 2022, almost a quarter of cars in Norway were plug-in electric vehicles, and there were over a million electric cars in the United Kingdom.[1] The shift from internal combustion engines towards electric motors has given new car manufacturers the opportunity to emerge and has greatly challenged existing manufacturers to adapt.

This development is also changing demand patterns. As the number of fossil-fuel-powered cars on our roads decreases, demand for resources such as lithium, nickel and cobalt, which are used in EV batteries, is on the rise.

What happens if supply cannot keep up with demand? In such a situation, we would depend on scientists to explore alternative materials for battery production. A few chemical compounds are known to be usable for this purpose, but they represent only a small fraction of the combinations that could create materials suitable for energy storage. Faced with time constraints and a vast array of potential avenues to explore, where should researchers begin?

Generally, scientists test new chemical combinations based on database searches and their own intuition. Testing each combination can be time-consuming. Some combinations can be quickly ruled out because their effects are already known or can be predicted with high accuracy, but this doesn't significantly reduce the vastness of the exploration area, often referred to by researchers as the 'search space'.

The discovery process can also lead to findings that are strikingly similar to existing materials and which rely on the same raw materials already used to create batteries. A scientist or a research group focusing on one type of battery chemistry might find it very hard to look at the task from a different perspective and come up with 'left field' combinations to test. This isn't ideal when the search is motivated by a shortage of raw materials in the first place.

Behavioural scientists can explain the phenomenon of inventors refining existing ideas rather than creating revolutionary ones. It is called availability bias. We rely on examples that readily come to mind when evaluating an idea. If you have a hammer in your hand, everything looks like a nail. Similarly, if a scientist has spent a significant portion of their academic career working on lithium-based batteries, their new ideas will likely be influenced by their knowledge of lithium.

Is it possible for algorithms to overcome the challenges of slow experimentation and availability bias? A team of researchers at the University of Liverpool thinks so. It assigned the task of finding promising new chemical combinations for energy storage to an AI algorithm it had developed, which uses knowledge of previously useful combinations to identify and rank promising new ones.[2] The researchers then conducted lab experiments on the combinations 'shortlisted' by the algorithm.

Remarkably, this led to four valuable discoveries, one of which included an entirely new family of solid-state materials, known as solid electrolytes, which play a crucial role in battery development. If a person had made this groundbreaking discovery, we might refer to them as a 'blue-sky thinker' or a 'left-field thinker.' However, we don't yet have a name for algorithms that demonstrate such innovative thinking.

According to Matt Rosseinsky, one of the Liverpool researchers, the collaboration between human and algorithm was key to the project's success: 'This collaborative approach combines the ability of computers to look at the relationships between several hundred thousand known materials, a scale unattainable for humans, and the expert knowledge and critical thinking of human researchers that leads to creative advances,' he explained.[3]

'This tool is an example of one of many collaborative artificial intelligence approaches likely to benefit scientists in the future,' Rosseinsky also said. The statement sounds perfectly reasonable, doesn't it? Not when you remember availability bias: when you're a scientist and develop an AI tool, of course you will think about how it helps scientists. Let me apply my own availability bias here too: I believe such collaborative AI approaches will benefit not only scientists but also entrepreneurs and creative individuals.

For large organisations, availability bias is not such a problem, and it's okay, even desirable, if new ideas are linked to what the business already does. They might like to completely reinvent themselves, but they are constrained by who they are and what they currently do. While they can use their market position and existing resources to their advantage, they do not have the luxury of acting like a startup: they can't quickly jump industries or drastically change what they do in response to customer feedback. The trick is in finding the balance between micro-innovations (slight improvements to current business models) and drastic shifts, which could benefit the business but also put its existence at risk.

Netflix is often used as an example of successful innovation because it reinvented the movie-rental space. It started as a mail-delivery DVD-rental service in 1997, charging customers per DVD rented. By 2000, it had introduced a flat-fee subscription model, offering unlimited rentals, without late-return or shipping fees. In 2007, it launched a video-streaming service, and in 2011 started to produce its own content, releasing its first original show, *House of Cards*, in 2013. But, arguably, the company's most exciting and unusual development only began in 2021, when Netflix began to expand into the gaming industry.

Why would a video-streaming and film-and-television production business enter the video-game sector? The answer is in its commitment to understanding customer needs and the ability of its sophisticated algorithms to understand customers much better than any of Netflix's competitors could. Its relentless curiosity about local audiences and its constant contact with and intimate knowledge of them has led to Netflix being a leader in creating original content that is attractive to local markets. How long will it remain a leader? Its expansion into the gaming industry might be Netflix's latest attempt to remain one step ahead of the competition.

To date, Netflix has acquired three game studios: Night School Studios, Boss Fight Entertainment and Next Games, which it bought in April 2022.[4] Next Games is a Finnish mobile-game developer that creates game experiences for TV and movie fans, building so-called franchise games. Franchise games are the sweet spot for Netflix: fans of its shows, such as *Stranger Things* and *Nailed It*, can now become part of the action. And more than 230 million paying users of Netflix (and another 100 million households that share the passwords of other users, according to Netflix) can now seamlessly switch between movies, shows and games when they arrive home late at night.[5]

Will Netflix enter other industries too? What about social networks? It seems far-fetched right now, but they're leaving the door ajar. Its corporate website identifies its competitors as 'all the activities that consumers have at their disposal in their leisure time. This includes watching content on other streaming services, linear TV, DVD or TVOD [transactional video on demand], but also reading a book, surfing YouTube, playing video games, socializing on Facebook, going out to dinner with friends or enjoying a glass of wine with their partner.'

In this section of the book, I want to explore a unique way of using algorithms – not for recommending actions or performing tasks for us, or for predicting the future, but for inspiring innovation and growth.

Successful companies in the age of algorithms are inherently curious. They prioritise asking questions above having all the answers, and they aren't afraid to venture into uncharted territories. They also experiment on a grand scale, and once they find a new solution to a challenge, they continue to seek out other ways to expand their market dominance.

Successful organisations aren't aiming for a sustainable competitive advantage anymore. They understand that sustained advantage is not possible in a moving world. Instead, they aim for transient advantage, a concept introduced by author and management professor Rita McGrath.[6] In the modern world, opportunities come up and disappear quickly. Any advantage that a business develops in the context of an opportunity is ... transient. The window of opportunity to respond soon shrinks. This calls for a move towards *continuous evolution*.

In this chapter, I'll discuss three strategies for putting organisations on the path of continuous evolution, and on their way to being successful RACERS. These strategies involve embracing digitalisation rather than just digitisation, building sandboxes for experimentation and nurturing a curious mindset. By seizing transient advantages and making the most of fleeting opportunities, businesses can stay ahead even as the landscape around them keeps changing.

Rule 4: Launch new value propositions

Sometimes the only way to tell whether a business idea is a good one is to test it in the wild. Case in point: over twenty years ago, Amazon opened up its internal applications to software developers from outside the organisation, launching its new product Amazon Web Services in 2002.[7] It allowed software developers to add Amazon product-search and shopping functions to their own websites through so-called web services. A web service is a tool that lets different applications talk to each other over the internet – in this case, Amazon applications and applications created by any developer in the world. Amazon fully automated the process: there were no human interactions required to gain access to its web services.

By 2004, over a hundred applications had been built thanks to the web services that Amazon offered. This wasn't a mind-blowing number, but Amazon was testing an idea, and the uptake was decent enough to convince executives that there was interest in using Amazon's infrastructure. This experiment gave Amazon the confidence to seize an opportunity that arose a couple of years later.

With the growing popularity of online shopping, the retailer soon needed to invest in server infrastructure. While everyday online retail can be pretty tame, it turns into a shopping frenzy a few times a year, and Amazon needed more powerful server infrastructure to support it. But the new infrastructure that would keep Amazon's services up and running during periods of heavy traffic would mostly stay unused for the rest of the year.

A few people in the organisation had a nifty idea for making the most of this under-utilised resource. Chris Pinkham and Benjamin

Black, two members of the Amazon IT infrastructure team, wrote a paper describing how Amazon could completely automate its infrastructure. Towards the end of the paper, they suggested that 'virtual servers' could be sold as a service, opening a window for Amazon to generate revenue from the infrastructure it had to build to support its retail business.[8]

Amazon decided to conduct a small experiment: it made some of its existing IT infrastructure available for paid public use. Pricing was based on the quality of the service customers received – servers that would always be available were more expensive than those that could 'go down' at times (when Amazon needed them during busy retail periods). The experiment was a success and led to the establishment of Amazon Web Services Inc. in 2006. In 2022 alone, AWS generated $80 billion in revenue.[9] It is now the largest cloud-services provider in the world.

Amazon made a move that others would later copy. It developed internal infrastructure to help the business operate smoothly and scale easily, and then it asked itself, *What is possible now that wasn't possible before?* This resulted in an entirely new value proposition: previously, Amazon had just sold products; now it was selling access to its own computing infrastructure. And there were customers happy to pay for it. Servers that would otherwise stay dormant became a source of income for the organisation. In corporate speak, the server division turned from a cost centre to a revenue centre.

Digital giants don't automate just for the sake of it; they automate to create new value and explore new markets, just as Amazon did. When working with businesses, I like to differentiate between digitisation and digitalisation. Digitisation involves leveraging technology to become faster, more efficient and cost-effective; digitalisation focuses on using these technologies to create distinct and unique value propositions. Digitalisation can give rise to spin-off businesses and present an organisation with significant opportunities to enter new markets and industries (all examples of new value propositions), making it more resilient to disruption. Not everyone grasps the difference between these two terms, but experts in the field understand it well – I didn't invent the distinction myself.

When an organisation has a well-defined and inflexible value proposition, it might choose to focus solely on digitisation. Digitisation is about finding better ways to achieve established value propositions. However, not many businesses can afford to be tied to a fixed value proposition.

Completing digitisation before starting digitalisation is not necessary. In fact, waiting is not recommended. Both digitisation and digitalisation are ongoing, potentially never-ending, processes, not discrete projects with definitive end points. Whenever digitisation is introduced it's always worth considering, *What is possible now that wasn't possible before? Does this automation create new opportunities or enhance existing ones?*

My colleagues and I studied thirty-two manufacturing companies from around the world that are known for their innovative success in Industry 4.0. Also known as the fourth industrial revolution, Industry 4.0 is the latest phase in the evolution of manufacturing and production, in which smart technology such as computers, robots and algorithms work together to make processes smoother and more efficient. Industry 4.0 focuses on the integration of digital technologies into factories and businesses, such as self-driving robots to manage inventory in warehouses and artificial intelligence to predict and prevent equipment failures in manufacturing plants.

Interestingly, the companies my colleagues and I studied aren't the disruptive forces that are often associated with Silicon Valley stories, but our research revealed that Industry 4.0 innovators set themselves apart by using ten key strategies that lead to the establishment of new value propositions.[10]

The first three revolve around enriching products or services, which could involve enabling lifelong partnerships, selling products as a service and charging for results rather than the product itself.

For instance, a modern car can communicate with its manufacturer and request service check-ups, making the purchase of the car just the beginning of the relationship between manufacturer and customer.

AVL List, an Austrian power-train manufacturer, has introduced remote usage and condition monitoring of the components it sells. Through the monitoring service, it establishes an ongoing communication channel between itself and the customer, allowing it to offer more products and services.

The market for yellow goods – construction and earthmoving equipment – is ripe for innovation in this space. Companies such as Finnish Konecranes Oyj, a crane manufacturer, are now offering cranes for rent, thus tapping into a new customer base. And German Kaeser Compressors Inc. has taken the innovative approach of selling compressed air by the cubic meter, only charging customers for the ultimate gain they receive from its products.

The next three strategies focus on reimagining processes: inviting anyone interested in collaboration to imagine new products, sharing infrastructure with others and moving from mass production to mass customisation.

American car manufacturer Local Motors opened up its product-development process by building a community of designers and technologists, called Launch Forth, to develop new products. And Adidas now leverages automation and robotisation to offer personalised shoes at a price similar to that of mass-produced sneakers.

The final four strategies capitalise on the knowledge and assets that organisations already possess – for instance, by selling product and transformation expertise, monetising collected data and turning in-house solutions into marketable products.

CLAAS, a German manufacturer of agricultural machinery, now uses data from its tractors to offer additional services through its spin-off, 365Farmnet. And when Orange Sky Australia, a charity providing mobile laundry and shower services to homeless people across the country, launched its volunteer management software platform, Campfire, it opened it up for other charities to use as well, under the new name Volaby.

A few years ago, I gave a TEDx talk: 'Creativity Not Required: How Great Minds Craft Ideas'.[11] In my speech, I shared the concept of structured ideation. Structured ideation is a problem-solving method that uses a systematic approach to generate innovative ideas. Instead of 'waiting for inspiration to strike', people using this method ask carefully curated questions to prompt themselves to explore new possibilities and uncover unique solutions. It's a valuable tool for discovering new ways to provide value to customers, the goal at the heart of digitalisation. By asking curated questions such as 'How would my local supermarket run my business?', you can generate some surprising ideas for value-propositions.

Rule 5: Start a digital evolution

Change in an organisation can be scary. But small experiments are possible for even the biggest, most change-resistant organisation, and they can alleviate some of that fear.

The state of Queensland in Australia's north is vast – if it were a country, it would be among the twenty largest in the world. You can imagine the challenges the state government has to deal with when launching new services.

The Department of Transport and Main Roads is in the process of introducing a digital driver's licence.[12] This smartphone application will allow drivers to leave their physical driver's licence at home if they want to. The idea is quite exciting: you will finally be able to ditch your wallet. And if someone needs to check your licence for certain details, you will be able to simply show them a QR code, choosing how much you wish to reveal: full details for the police, and perhaps only confirmation that you're over eighteen (no date of birth, home address or other private information) for nightclub bouncers.

But can it work in Queensland? What happens if a driver needs to show someone their licence but they're in an area where there's no phone reception? Will the police be able to scan the QR code displayed by the app if the owner's phone screen is shattered?

To address these issues, the department ran pilot tests in two cities. The trial revealed that many young people have phones with cracked screens, making QR codes unreadable. As a result, the team needed to find an alternative way for licences to be checked. The trial also highlighted some unexpected challenges, such as businesses that wanted to photocopy driver licences. To accommodate this, the app now offers a printable version of the licence that can be emailed or texted.

The trial taught the transport department some valuable lessons and helped it to limit the impact of potential problems. While it might sound like an obvious idea to run small pilot studies such as these before launching a product, not a lot of other public-sector organisations are even considering it.

Now, contrast the transport department's experiment with the experimentation that automation allows businesses to achieve: hundreds or even thousands of experiments can be run at any given point in time. In a sensitive scenario, the scope of experimentation might be constrained by legislation, but otherwise there are plenty of opportunities to experiment at scale. For instance, a business that automatically generates the notices it sends to customers could experiment with slightly modifying every single document it sends. If the business properly designs such experiments – good data scientists are indispensable here – it can lead to some great insights.

Former US president Barack Obama used the power of such large-scale experiments to increase his campaign fundraising before the 2008 presidential election.[13] His team's approach was simple yet powerful. It focused on the online presence of Obama. Visitors to his webpage were met with a colourful image of Obama and a large 'Sign up' button. The team decided to experiment: Would a different choice of text get more people to join the campaign and donate? What about the image? Could changing the photo have an impact on conversion rates?

When thousands of people visit a website, and the algorithm behind it is capable of showing a slightly different version to each visitor, it is possible to run large numbers of experiments simultaneously. The team was keen to try this approach.

Some visitors saw 'Join Us Now', instead of 'Sign up'. Others saw 'Learn More'. The team's analysis showed that the number of sign-ups was almost a fifth higher when visitors saw 'Learn More', as opposed to the original 'Sign up'. And the photo? It turned out that a black-and-white photo of the Obama family increased sign-ups by slightly over 13 per cent. When both the black-and-white photo and the 'Learn More' text were shown to visitors, sign-ups increased by 40 per cent.

During the campaign, thirteen million people signed up to support Obama. Four million of them can be attributed to these experiments. And about $75 million in money raised is believed to be the result of improvements introduced by this digital evolution.

Running a business can sometimes feel like Groundhog Day. Sure, almost every day comes with new and unexpected challenges, but if you zoom out a tiny bit it all looks similar. At some stage, processes might become repetitive. Many of us face this challenge in our personal lives too. Do you always get up at the same time? Do you always have your lunch at the same place? Getting stuck in a rut, whether in life or in business, can be detrimental.

When I first watched Max Hawkins' TEDx speech, I was confused. Hawkins, an artist and computer scientist, used the TEDx platform to share how he had given over control of his life to an algorithm.[14] Prior to doing this, Hawkins had optimised every aspect of his life: his wake-up time, his coffee-shop choice, his commute. He was a software engineer at Google, and this was the mindset to have. It also was the mindset of almost every other software engineer in Silicon Valley. One day, when Hawkins realised how predictable his life had become, he decided to write an algorithm to help him break away from the monotonous, repetitive reality he had created for himself.

Hawkins started by writing an application that would help him explore new places on Friday nights. Instead of going to the usual bars or restaurants, it would send him and his friends to a random place in San Francisco that was listed on Google Maps. The app would order an Uber, telling the driver where to go but not Max and his friends. Wherever they arrived, it would be a surprise. The first time Hawkins used his app, it took him and his friend Kelly to Psychiatric Emergency Services at Zuckerberg San Francisco General Hospital and Trauma Center. Who said algorithms don't have a sense of humour?

Hawkins was hooked. He used the app to explore random places all over town. 'I started discovering that there was an entire side to San Francisco that I had been ignoring because of my preference,' he told his TEDx audience.[15]

He then gave up even more control to the algorithm, allowing it to make random choices for him about food, music, plans for the day and even tattoos.[16] At the more extreme end, he moved to other parts of the world based on the algorithm's decisions.

My initial reaction was that Hawkins' experiment was immature and downright risky. But when I thought more about it, I realised Hawkins was demonstrating the potential benefits of breaking out of the Groundhog Day rhythm. Hawkins had decided to use an algorithm to bring variety into his life.

Even if Hawkins slightly exaggerated the amount of control he allowed the algorithm to have – I am sure he ignored suggestions that were dangerous or plain stupid – he had created a very structured system that introduced random mutations into his life.[17] He reports that, paradoxically, giving up this control gave him a sense of freedom:

> I discovered that my preference had blinded me [to] the complexity and the richness of the world. And following the computer gave me the courage to live outside of my comfort zone, to discover parts of the human experience that I ignored because they were too different or not for me.[18]

But evolution is not just about random mutations. It is also about the survival of the fittest. Those mutations that improve the so-called fitness function are meant to stay. (In animal life the fitness function would be survival, in business it might be revenue and in daily life it could be happiness.) While Hawkins didn't say which of the application's random ideas were one-offs and which became part of his life, he did say that his life choices changed as a result.

I hope by now you can see how my perception of algorithmic randomness in life has completely shifted. I went from being shocked by the Hawkins' experiment to seeing how similar experiments could make our lives – and our businesses – richer.

Taking inspiration from Hawkins, I began to incorporate random algorithmic choices into my work. In the MBA courses I teach, I allow 20 per cent of each session's content to be randomly generated. I've found this to be a safe balance. Even if the random ideas aren't that helpful to students, they only make up a small portion of the session, and more often than not they lead to engaging discussions, making each lecture unique.

I also use this approach in conference presentations. At one event, I increased the randomness to 80 per cent, leaving only 20 per cent under my control. I discuss this method further in my sixth rule, Stay Curious.

In some cases, businesses don't even have to consciously experiment. An experiment has already occurred, and they just need to identify it as such.

At 2 a.m. on 11 October 2021, David Card's phone started buzzing. Card, a professor of economics at the University of California,

Berkeley, thought it was a prank call from one of his friends, Tim.[19] But something didn't add up: the area code of the caller showed as '+46'. The call was coming from Sweden. Card answered his phone to receive probably the most stunning news on his life: together with two of his colleagues, Joshua Angrist and Guido Imbens, he had won the Nobel Prize in Economic Sciences.

Card, Angrist and Imbens study labour markets. Most researchers either analyse the data they have access to or they create experiments, introducing interventions and then testing the impact of those interventions. Simply speaking, they either analyse the past or try to shape the future. Scale and ethics are common challenges associated with experiments. It is expensive to run large experiments, and some of them could, in practice, be harmful, exploitative or destructive. Sound harsh? Imagine trying to understand how people cope with financial stress by intentionally leading an individual or a group of people into bankruptcy.

But Card and his colleagues devised a different approach. They observed that the natural randomness of the world can resemble the randomness that scientists introduce into their experiments to ensure the validity of their results. In short, Card and his colleagues showed that it is possible to draw scientifically valid conclusions from 'random experiments' that have naturally occurred in the past. Rather than manipulating someone into bankruptcy, it is possible to identify people who are already bankrupt and then study their coping mechanisms.

How can businesses tap into this Nobel Prize–worthy insight? We are yet to see the business community adopt this approach broadly, but some examples are emerging. My favourite academic study[20] – it's very academic[21] – looks at the behaviour of over a million people who ran more than 350 million kilometres over five years. Without conducting a single experiment, the researchers were able to show that when less active runners run farther or for a longer time, it motivates their friends who are more active runners to also run farther or longer, but it doesn't work the other way round. They also showed that male runners are influenced by both male and female runners, but female runners are only influenced by other women.

Insights such as these could be applied immediately by brand ambassadors, advertisers or online running applications. For instance, if a running network wants its female users to run more often, and therefore use its application more, it should alert female users to the running activity of their female friends. Alerting them to the progress made by their male friends would be a waste of 'feed space'.

To grow and evolve, sometimes all you need to do is look into the past.

Rule 6: Stay curious

Can you recall how many different products Google has launched over the years? Ten? Twenty? Fifty? If you're not sure, you can consult a website called Killed by Google, which lists all of the products Google has launched and then taken down.[22] The list of its aborted experiments is currently 289 lines long. It includes the social network Google Plus and its lesser-known predecessor, Orkut. And, of course, the list keeps growing. But the nearly three hundred projects that Google has introduced and then killed off are not failures. They are the natural outcome of evolutionary exploration – a digital survival of the fittest.

We don't give businesses such as Google enough credit for this exploration. It's easy to point out that its search engine and advertising revenue are still Google's main cash cow (over 80 per cent of its revenue comes from advertising alone).[23] But Google, probably recognising the fragility of its business model, keeps asking itself, *What else could our customers pay for?*

Whatever its motivation is, Google is curious.

When managing a successful business, it might seem like coming up with new questions to ask could be a distraction. Companies often concentrate on answering questions instead, and they usually have a standard set of queries. They seek solutions to questions such as 'How much should we charge for our products?', 'How can we improve customer satisfaction?' and 'How can we reduce our tax liabilities?' They also explore hypotheses and 'what-if' scenarios, such as 'What would happen if we stopped advertising our products?' and 'If we change the packaging colour to yellow, will customers notice our product more easily and buy more of it?'

It appears that each high-level executive position comes with a specific set of questions to address. The Chief Marketing

Officer focuses on product-related questions, the Chief Financial Officer deals with budget-related questions and the CEO asks themselves how to make the business attractive to investors, especially if it is a publicly traded company. This *analytical curiosity*, which involves proposing hypotheses and testing solutions, has been around for a long time and is taught to students by business schools.

But there are also companies that focus on discovering new questions instead of only answering predefined ones. By allowing their curiosity to lead them, businesses can explore uncharted territory and pose questions no one else has considered before. Such an innovative mindset can be incredibly powerful in driving growth and differentiation within industries.

One example of a company that adopted this approach is The DAO. Its founders asked themselves whether a business could exist without humans in charge. Similarly, the creators of Avatar Robot Cafe DAWN ver.β questioned whether it was possible to introduce robots into their restaurant while maintaining, or even enhancing, a welcoming atmosphere. By daring to ask unconventional questions, these companies have opened doors to entirely new business models.

Such *design curiosity* thrives on experiences and emotions, and seeks to identify problems or opportunities in a variety of situations. Design-curious organisations look beyond the traditional boundaries of their industry and embrace novel ideas. Although the approach is relatively rare in the organisations I have encountered, those I have seen apply such curiosity often break free from their industry's norms, either reimagining their industry or venturing into an entirely new one.

Cultivating design curiosity can lead to a more adaptable and forward-thinking organisation with a strong culture of innovation that allows it to stay ahead of the curve. By continuously seeking new questions to ask and embracing the unknown, such businesses can unlock their full potential and achieve remarkable success.

And then there is *generative curiosity*, a unique approach to problem-solving and innovation in the business world, made possible by rapid advancements in technology and the increasing prevalence of algorithms. In short, algorithms are let loose to create – or generate – questions. Just as Hawkins' algorithm asked, 'What would happen if you dined in this randomly chosen, restaurant tonight?' Unlike other forms of curiosity, generative curiosity is unconstrained by traditional boundaries or specific sets of questions. Instead, it emphasises the power of random exploration and experimentation to achieve positive business outcomes. This exciting approach has the potential to revolutionise the way businesses innovate and adapt to changing market conditions.

An essential aspect of generative curiosity is its relationship to failure. It sees failure not as a negative outcome but as an invaluable learning experience. In fact, without failure there can be no generativity. By embracing the notion that failure is an inevitable part of the innovation process, businesses can learn from their missteps, rework their ideas and ultimately achieve greater success in the long run.

Companies that embrace failure and harness the power of random exploration can discover new opportunities and drive innovation like never before, developing groundbreaking products, services and business models, creating lasting value for themselves and their customers, and staying ahead of the competition.

In 2022, I decided to use the generative-curiosity approach myself. I was invited to give a presentation at a conference that brought together the principals of Queensland's independent public schools. There are 250 such schools in the state. Their principals meet regularly, and they reach out to me every few years to learn more about what impact the latest technology could have on their work. This time I was asked to share my thoughts about artificial intelligence and schools.

I am not particularly keen on preparing presentations. Some would call me lazy. I don't mind sharing my thoughts on stage, but I absolutely hate coming up with titles, slides and abstracts. Out of this laziness, I thought to myself, 'Perhaps I could get an AI to write the title and abstract for me.'

As you may remember, many modern AI systems have natural language interfaces. This means we can simply tell them what we need. Choosing one, I wrote the following prompt for it: 'Write a clickbait title for a presentation on the impact of artificial intelligence in education. The presentation, delivered by Prof. Marek Kowalkiewicz, will be shown to over a hundred school principals.'[24]

Here is how the algorithm responded: 'How to Beat the Robots: A Guide to Ensure Your Students Stay Ahead in the Age of AI.'

Encouraged by these results, I also asked the algorithm to generate an abstract. Once again, I wrote a prompt: 'Write a two-sentence abstract of a presentation on the impact of artificial intelligence on education and the future of our children. Make the first sentence a clickbait question.'

The algorithm's response was pretty good: 'Is artificial intelligence the end of education as we know it? In this presentation, we'll explore the impact of AI on education and how it will shape the future of our children.'

I sent the title and abstract to the organisers, who added them to the agenda. They thanked me for proposing such an engaging session and for my thoughtfulness in specifically tailoring it for school leaders.

This was also encouraging, but I still felt quite lazy. And that's when it struck me: I could get the algorithm to write the slides for me too! By that stage, I'd become just like Max Hawkins, ready to submit to the will of the algorithm. I felt excitement! It was a twenty-minute presentation, and I asked the algorithm to prepare ten slide titles.[25]

It took some more time to create nice images for the slides: I used another algorithm to do that, simply providing it with the slide titles and nothing else. I decided not to add any other text to the slides in addition to the titles.

When the day came for me to deliver my presentation, I went on stage. Without telling anyone about it up-front, I delivered a presentation that had been entirely scripted and illustrated by algorithms. It was not easy at all: while the process of generating the slides took almost no effort, delivering a compelling presentation was much harder. It was a bit like presenting somebody else's slides: I had to find a way to tie together all of the algorithms' ideas, I had to add stories to the slides and – this was particularly hard – I needed to create smooth transitions between each slide. When I finished presenting the ten slides, the additional mental load it had required left me internally drained.

When the audience clapped and I saw that the emcee was about to walk on stage to thank me, I pointed out that I still had a few minutes left. I moved on to the second part of my presentation, which was entirely mine. I disclosed, to a stunned audience, that the previous ten slides had been produced by algorithms in their entirety. I walked the audience through the process, explaining the capabilities of modern AI algorithms as I went. I vividly recall asking how many people in the audience were aware of GPT-3: only one or two of the school principals raised their hands. This was not surprising – ChatGPT hadn't been released yet. I also recall asking every single person in the room to add 'read more about GPT-3' to their post-conference homework. Little did we know that ChatGPT would shake the education sector to its core just a few months later.

After the conference, the principals wrote to me with their thoughts. One said, 'The AI certainly covered the information, but you brought it to life!' This highlights the importance of working with algorithms as partners. Another wrote: 'Mind blowing! Maybe I need to work smarter . . .' This principal was probably referring to a 'work smarter, not harder' line I must have used in the presentation. Indeed, as these tools become more and more capable, they will allow us to focus on the tasks that make us stand out.

But will I repeat my experiment? Likely not, and definitely not if a presentation really matters to me. While the quality of the content generated by the algorithm was impressive, the originality of its insights was not impressive at all. Regurgitated ideas – which is what a lot of text-generating algorithms produce – are a time-filler, nothing else. They don't bring much to the discussion, only repeating and potentially reinforcing points that should already be obvious.

Will I use algorithms to help me prepare for future presentations? Yes, without question. But in a very different way. Instead of using an algorithm to generate ten slide titles, I will ask the system to generate hundreds, and perhaps thousands, of ideas. Then I'll sift through these ideas, mixing them up and rearranging them, to find one or two that stand out. Using an algorithm to feed my curiosity would be a better use for it, and I would definitely appreciate this type of inspiration.

In his 1962 book *Profiles of the Future*,[26] British science-fiction writer Arthur C. Clarke offers three observations, often referred to as Clarke's three laws. His third law is popular and widely cited: 'Any sufficiently advanced technology is indistinguishable from magic.' His first law is less known but very accurate: 'When a distinguished but elderly scientist states that something is possible, they are almost certainly right. When they state that something is impossible, they are very probably wrong.' But it is Clarke's second law that I find the most inspiring: 'The only way of discovering the limits of the possible is to venture a little way past them into the impossible.' These 'unknowns', just a little way past the realm of the 'possible', can serve as a great motivation for our curiosity – both in life and in business.

Can such curiosity be codified and exhibited by algorithms? Apparently, yes. You might remember the definition of curiosity – taken from the field of AI – that I used earlier in the book: curiosity is 'the error in an agent's ability to predict the consequence of its own actions'.[27] Algorithmic curiosity emerges when an algorithm has no idea what effect an action might have. If the algorithm decides to take this action to find out what will happen, that's an act driven by curiosity.

Without such curiosity, you will never wander off the beaten track. In 2021, the Queensland University of Technology, together with the Massachusetts Institute of Technology, launched a series of webinars on the attributes of the future enterprise.[28] 'Future enterprise' is a term my colleagues and I coined to describe the exploration of innovative business models, technologies and attributes shaping the business entities (not necessarily corporations) of the future. I was invited to speak at one of the webinars and was joined by Professor Steven Eppinger of MIT. Eppinger has a background in engineering and project management, while mine is in computer science and management. We discussed curiosity as an attribute of a future enterprise and shared our experiences working with various organisations to hone their curiosity.

We had both found that successful organisations often embark on aimless, but not pointless, exploration. We also concluded that a little process goes a long way – combining design thinking and agility is essential. Finally, we stressed the importance of ongoing reflection in helping organisations learn from their exploration and refine how they operate – their strategies, processes and value propositions.

Our overall message was simple: curiosity is all about exploring unknowns, and algorithms can help us with that. Just like Max Hawkins, who allowed algorithms to take him to unknown places, any business or individual can ask an algorithm to show them these unknowns. Will it lead to better business or better life experiences? We have no idea. But that's exactly the point.

Children love exploring the world. They ask questions, learn from experience and are always picking up new skills. What drives them to do it? It's not the promise of a reward – they don't yet care about salary or profit, status symbols or peer perception. Their behaviour is fuelled by an innate drive: curiosity. Their curiosity exposes them to new experiences that might help them later in life.

Time and time again, we learn that curiosity benefits business too. In fact, many startups – the equivalent of children in the business world – are hypothesis-driven. They ask questions they don't know the answers to – *Would customers buy this product?* – and embark on exploring the unknown. Thanks to the parent-like support of venture capitalists, they don't yet care about profits either. Like children, they don't just ask questions, they also learn from experience, picking up the new skills that they will need to succeed later on.

The idea that algorithms in the economy could be driven by the same type of behaviour is counterintuitive – we expect algorithms to behave predictably. Computer scientists use the term 'deterministic' to describe such behaviour: if the input of an algorithm is known, it is possible to calculate – or determine – what the output is going to be in every case. But such thinking, as we have seen in this chapter, is incorrect. Only some algorithms are deterministic. Most algorithms, especially the ones this book focuses on, can and do introduce randomness. They exhibit non-deterministic behaviour. And this is surprisingly effective at introducing novel ideas, especially if done at scale. Perhaps we overestimate the human ability to come up with great ideas if random variations introduced by algorithms can produce outcomes better than those humans can think up.

The AI researchers who proposed the definition of curiosity I cited earlier in this chapter[29] made curiosity an innate drive – a reward function, to use the proper term – of an algorithm they asked to play two computer games, *VizDoom* and *Super Mario Bros*. *VizDoom* is a version of *Doom*, a first-person shooter game from the 1990s, in which the player navigates a space base while trying to fend off attacks from monsters[30]. *VizDoom* was created solely for the purpose of testing AI algorithms. *Super Mario Bros.* is a so-called platform game, in which the player controls the main character, Mario, and attempts to move through a series of side-scrolling levels by surviving various hazards, defeating enemies and collecting rewards.

When the researchers made curiosity a 'driver' of algorithmic action, the algorithms made very interesting progress.

In *VizDoom*, the algorithm started exploring the space base. It didn't just randomly walk in all directions: its actions had to have some purpose or it would likely have gotten stuck in one place for too long, just like a fly randomly buzzing around a room. The algorithm explored the entire base, satisfying its curiosity.[31] It wasn't given any 'rewards' by the game – there were no extra points for walking around. Curiosity was its only driver.

In *Super Mario Bros.*, the algorithm learned how to play a significant part of one level. It didn't need any other incentives to make progress. Just like the algorithm playing *VizDoom*, it was motivated by curiosity alone. But then it got stuck. There is an obstacle in one section of the game – a pit – that can only be crossed if the player presses a sequence of between fifteen and twenty keys on the game controller. The curiosity-driven algorithm was unable to make it across the pit – that feat could only be achieved by an execution-driven Mario. And because there was no execution-driven Mario available to take over the controls for a while, it was game over.

What a metaphor for the business world.

9

Be Boldly Optimistic: Relationship Saturation

In 2014, Facebook CEO Mark Zuckerberg said he wanted to create a 'dial tone' for the internet[1] – that is, a universal access point for cyberspace. In the same way that anyone can pick up a landline phone and hear a dial tone, indicating the phone line is ready to be used, people would have free access to a set of basic services, such as messaging and weather forecasts, and a portal for connecting to the internet. He wanted Facebook to be used almost like a utility, a fundamental part of the daily internet experience. His broader goal was for every person on the planet to use Facebook.

Just a couple of months earlier, Facebook had reported that its user base was 1.23 billion. As I am writing this book, it's nearing three billion users: that's almost 40 per cent of the population of our planet, and about 50 per cent if we exclude children younger than fourteen years old, who shouldn't be using Facebook anyway.

Say what you want about Facebook – there are countless reasons to stay away from it, including its complete disregard for users' privacy, its lack of moral compass and its inability to come up with alternative business models – but there's no denying it knows how to set ambitious goals.

What happens when a business has so many customers that it literally runs out of non-customers, as Facebook might soon do? Not every organisation sees the entire globe as its market: the entire so-called 'addressable' market can be orders of magnitude smaller for a business without global ambitions. What if it still wants to grow?

The only solution is to offer more to *existing* customers. Businesses call this practice 'cross-selling'.

Algorithms can help with cross-selling. They don't sleep or take too many breaks to recharge, and they're almost never late.[2] They're also quite portable: they can run on a mobile phone, inside a car, in the cloud or on any device imaginable. This makes it possible for the businesses and individuals who control these algorithms to connect them with customers.

This phenomenon – the omnipresence of algorithm-enabled technology – is known by scientists as 'ubiquitous computing', a term coined in the late 1980s.[3] This is quite common right now. In all likelihood, if you look at your current surroundings, you will be able to point to a surprisingly large number of small 'computers' hidden in everyday items. You might have a smart speaker on your desk or bedside table, a smart lightbulb that can be controlled from your phone, a smart watch, a tiny computer controlling your microwave or TV, and perhaps even a lock on your front door that you can open by typing a code into a little keyboard displayed on a small screen. There are also computers in your washing machine and your smoke alarm. There might be a tiny computer in your doorbell. Many of these computers can communicate with each other, and some of them can run algorithms the way your smartphone can run apps.

Ubiquitous computing[4] is one of the keys to cross-selling. I prefer to call it attention saturation. Traditionally, businesses have aimed for *market saturation*, which is achieved by serving as many customers as possible. Market saturation is a scale game: businesses either do whatever they can to attract those customers they're not serving but could be or they consciously decide not to focus on those customers. Attention saturation goes deeper than that. Businesses can now ask, *Are there any other aspects of customers' lives where we could be of value?* And, again, businesses might decide some aspects of their customers' lives are not of interest.

In the past, businesses didn't have the luxury of considering such strategies. Attention saturation was simply impossible. At least at the scale that is possible right now.[5] Regardless of whether the addressable market of a business is a thousand customers or a billion, businesses can tap into the potential of algorithms to 'saturate' their relationships with customers.

But all businesses should be aware that there are lines that we would prefer they didn't cross. Let's explore an example of a business that comes pretty close to such a line.

A few years ago, my team at the Queensland University of Technology partnered with a large water-utility organisation. It came to us with a challenge to overcome. Its customers only heard from the company four times a year, when they received their quarterly invoices for water and sewage services. So far, so good, you might think. But let's be frank, no one loves opening bills. And when customers got in touch themselves, it was almost never because they were delighted with the service and wanted to share their happiness. They contacted the organisation to cancel or move their account, or to report an issue such as low water pressure, discoloured tap water, a burst pipe or overflowing sewage. Hardly moments to celebrate.

While the organisation is constrained geographically by where its pipes are, limiting the number of customers it can have, it realised its existing customers were underutilised. The organisation wanted to create more opportunities to interact with customers, which it hoped would give it a chance to provide value and therefore increase customer satisfaction.

My team ran a full-day 'design jam', in which employees from the organisation and QUT students came up with several ideas for business models. Their goal was to increase the frequency of customer interactions. We asked ourselves, *What if we could interact with customers daily instead of quarterly? How could we make that happen?*

One of the solutions suggested was a smart diagnostic toilet. The rationale was that if the organisation could collect more information about customers, it could proactively reach out to them to offer additional value. The initial concept was very simple: Customers would be able to install a small sensor inside their toilet bowl. Once installed, the sensor would start capturing information. Even the most basic information possible, such as how often the toilet was flushed, could lead to interesting insights, including whether anyone was at home and how many people, roughly, were there. This might not sound like very valuable information, but, hypothetically, it could be used for a number of purposes, such as alerting caregivers if their elderly parents hadn't used the toilet in a long time, or determining how many people were staying in a rental property. This is where it gets close to that line we'd prefer businesses to steer clear of.

Ideas are cheap, as the saying goes, and it's all in the execution. The general concept was picked up by a team of Executive MBA students at QUT, who further developed the business model, outlined the most acceptable business cases and proposed a business.

Their business design for the Smart Toilet Company – or Smart TC (they voted against having the word 'toilet' on their business cards) – won the 2016 Global Business Challenge,[6] a global graduate business-case competition.[7] The win came with a seed fund and strong support from a health department, which was keen to deploy the proposed technology in regional areas to understand community health.

Unfortunately, Smart TC didn't manage to build a sensor that could be produced at scale. However, the concept of having algorithms in your toilet has been picked up by other organisations. Toilet manufacturer Kohler released a smart toilet in 2019,[8] which includes a built-in Alexa smart assistant that allows users to order toilet paper (among other things) without getting up from their throne.[9] A Duke University spin-off, Coprata Inc., is now building smart toilets that can pack human waste for laboratory analysis and includes 'automatic tracking of bowel-movement characteristics'.[10]

Although it hasn't implemented the smart-toilet idea yet, the water-utility business I worked with did end up finding some effective ways to increase market saturation – and our design jam demonstrated the type of bold and optimistic thinking that the economy of algorithms makes possible.

What happens if your bold ambitions can't be achieved by you alone? Once again, algorithms can help. Just as software applications can share data and run coordinated processes within an organisation, algorithms can bring organisations together in a coordinated network.

If you've ever bought a plane ticket through a website offering low fares, you have already experienced the potential of such automated coordination. Even though the purchasing process often requires you to interact with a business completely separate from the airline (such as an online ticket reseller), the ticket is issued instantly, and immediately after making the purchase you can visit the airline's website to choose your seat and a meal suited to your dietary requirements. The integration is so good that even though at least two organisations are involved in the transaction, there is no perceptible delay. Some of us will remember how booking flights used to work. When tickets were purchased through a travel agent, the process took much longer – minutes, sometimes hours – and you had to wait for days before the tickets were 'confirmed'. The economy of algorithms has changed all that.

Rule 7: Maximise customer value

The world of algorithms has opened up new possibilities for businesses to strengthen their connections with customers, through smartphone apps, smart speakers and other devices. Even my smoke alarms can now communicate with me, and they do so on occasion. Recently, a team of builders working on a renovation at my house created enough dust to set off the alarm, and I received a smartphone notification while I was at work.

We wouldn't expect the smoke-alarm industry to think too much about saturating its relationships with the owners of smoke alarms. But you could argue that businesses in other industries would seriously consider using smoke alarms as an opportunity to further saturate *their* customer relationships. Case in point – my smoke alarms were made by Google, which is slowly expanding into manufacturing products that are used in every aspect of our lives.

Every interaction with a customer gives a business an opportunity to provide the customer with additional value and offer them new services and products. In the case of my smoke alarm, I needed to download an app onto my phone to configure and monitor the alarm, and this app is a portal to all sorts of 'smart home' services offered by the manufacturer.

In a June 2018 interview for *The Guardian*, Netflix CEO Reed Hastings denied the company's biggest competitor was HBO, Amazon Video or YouTube. 'When you watch a show from Netflix, and you get addicted to it, you stay up late at night,' he pointed out. 'We are competing with sleep.'[11] In another interview, he claimed that Netflix competes with red wine, something else people use to relax after a tiring day.[12,13]

Hastings understood that relaxation is one of the main reasons Netflix customers use the company's product – relaxation is one of their 'jobs to be done'. This understanding not only helps businesses know their customers better and improve their products and services but it also allows them to identify their competitors by prompting them to ask, *Who might get the job done better?* That's why Netflix looks to other industries – including gaming, social networks and even bars and distilleries – for ideas on how to maximise customer value and attract customers that would otherwise use very different products to achieve relaxation.

In some of the executive education sessions I have led, we've explored how a skatepark could compete with a fast-food restaurant for the attention of bored children, just as a masseuse might compete with a mattress store if the customer's job to be done is to relieve back pain.

Clayton Christensen, the late Harvard Business School professor who developed the 'jobs to be done' theory,[14] once ran a study outside a fast-food restaurant. He asked people who had ordered take-away milkshakes what jobs they had 'hired' the milkshakes for. It turned out most of them faced a boring commute, needed something to do with their hands and knew they would soon be hungry if they didn't eat or drink something. The job to be done by the milkshakes was to be a 'food companion' for commuters and to keep them from feeling hungry for a bit longer. The immediate outcome of the study was the realisation that while fruit, bagels and chocolate bars were competing with milkshakes, they weren't good substitutes for the particular job to be done. The long-lasting impact of the study was that it opened our minds to the idea that products might satisfy customer needs in unexpected ways.

I like to practise my jobs-to-be-done muscles whenever my children ask me for something. It's a fun exercise. For example, my son, who is thirteen, regularly wants me to take him to a fast-food restaurant, but on each occasion the job to be done might be different. Sometimes he's hungry, and a sandwich at home is a reasonable alternative. Sometimes he is bored, and we can play a game together instead. Sometimes he wants the free toys that come with the meals. I know you might think buying better-quality toys is an alternative, but no – in such instances, I wait it out.

American industrialist Henry Ford is often quoted as saying that if he had asked his customers what they wanted, they would have said they wanted a faster horse.[15] According to this quote, customers don't know what they want. It is therefore essential for businesses to focus on understanding the actual needs of customers. Consider customers who are in the market for a new car: some might want to make their neighbour jealous, others might want to get to work faster, and others might require a new vehicle for business. In each case, the perfect product or service will be different.

This way of thinking presents a massive opportunity for organisations that try to saturate their relationships, as it can allow them to identify other products or services that customers might need. This is where the economy of algorithms comes in: by opening up the possibility of connecting with customers 24/7, it creates so many opportunities to offer them new types of value.

We already know that algorithms are fast. They can make interactions feel instant. But what if this were taken further, and they were able to meet customers' needs even more quickly? There is a name for this approach: proactive service delivery. Organisations that deliver proactive services are called proactive organisations. What's so special about them? They know so much about their customers they can offer products and services the moment the need for them arises, often before the customer realises there is a need.

The Smart Toilet Company was set up with proactive service delivery in mind. The toilet used data collected by its sensors to understand whether users had any potential health issues, allowing the company to offer pre-emptive health services. A similar approach, slightly outside the domain of humans and toilets, is called predictive maintenance. By installing a small sensor on a piece of equipment – say, on a mining-truck engine – unusual vibrations can be detected and owners can be alerted to issues before they start causing problems.

There are significant privacy concerns associated with deploying proactive services. Of course, the two examples I've just discussed present very different concerns: installing a sensor inside a toilet bowl feels much more intrusive than putting one on an engine. But, regardless of the context, any proactive approach needs to consider data privacy as a priority.

So far, I have written about how the economy of algorithms helps businesses to understand customer needs and to proactively offer products and services of value. But this can be taken even further again: businesses can now offer customers products or services uniquely tailored to their needs.

One of the most extreme examples of personalisation is when a business considers every customer down to the level of their genetics. InsideTracker, a startup founded in 2009, offers personalised nutrition plans for its customers. In the words of its founders, the startup's mission is to 'help people add years to their lives and life to their years by optimising their bodies from the inside out'. It's a lofty goal, but the team behind the business consists of experts in genetics, biology and aging, whose insights are shared in peer-reviewed journals[16] – there is reason to believe they know what they're doing. Their system, an AI 'engine', makes diet and exercise recommendations based on information from blood tests, DNA analysis and activity trackers such as smart watches. Every customer sees results that are 'just for them'.

There is another famous quote attributed to Henry Ford: 'Any customer can have a car painted any color that he wants so long as it is black.'[17] This was the premise of the Industrial Revolution: machines allowed for production at scale, but every item produced was identical. Variation was too costly, and against the principles of industrialisation. The digital revolution changed that. Today, InsideTracker can have a million customers, and each will have a very different nutrition plan. Facebook can have billions of customers, and each will see a uniquely different timeline.

InsideTracker recently received another multimillion-dollar investment.[18] There are many investors who strongly believe in the power of mass personalisation, enabled by very fast, capable and autonomous algorithms.

Rule 8: Build digital ecosystems

Apple became a leading creator of smartphones and computers by coordinating a complex ecosystem of suppliers and manufacturers. Its phones are designed in California and built in China, Vietnam and India by Apple's partner, Foxconn, using parts made in countries such as the United States, Japan, Korea, Taiwan, the Czech Republic and Mongolia.[19] Manufacturers work with suppliers to produce the phones, which are then shipped worldwide by shipping companies and traded by a network of retailers. Thanks to the skilful coordination of the supply chain, they then end up in the hands of customers, who simply can't wait for the new model every year.

When you look at purely digital services, you'll notice such orchestration is even more common. Large retail sites combine online payment services, 'buy now, pay later' options and various delivery modes – all the while providing a consistent experience for customers. Remember the all-in-one (AIO) shopping bots I wrote about earlier in the book? They connect to multiple retail sites just to make sure you receive the product you want.

It seems that the most successful businesses in the economy of algorithms are those that coordinate broader ecosystems. Fortunately, opportunities for orchestration are not just available to Silicon Valley startups. Even the most traditional industries can benefit from it.

A few years ago, I worked with one of Australia's oldest businesses: a timber-manufacturing company. The company's expertise was in providing timber and some related design services, but it recognised that the customer's 'job to be done' went beyond just buying wood. It understood that the product customers bought was just one of many puzzle pieces they needed to combine to achieve what they really wanted. For instance, a customer might need to build a granny flat. To achieve this, they would need the timber, but not the timber alone. Once the business realised this, it designed its own orchestration services. It can now help its customers get council approval, find builders and perform a range of other activities associated with building a granny flat. This allowed the company to leverage its core competencies to create more value for its customers.

Other businesses in the economy of algorithms can also orchestrate value for their customers by creating a network of partners and services that add value to jobs to be done. In doing so, businesses can create a one-stop shop for the customer and become a trusted service provider, thereby creating a 'sticky' relationship with customers. Additionally, businesses become more efficient when they can offload certain activities to their partners.

The power of the economy of algorithms is in making such interactions with other businesses much easier – and in many cases, they can be fully automated.

There is a type of technology that helps with such business integrations. It is called 'application programming interfaces'. Most software developers see it as a way of connecting one application to another, but APIs can also connect businesses. I am not the only one to recognise the business potential of APIs: 'API economy' is a well-established term.

Anyone who buys a plane ticket these days benefits from such digital interfaces. There used to be a time when airlines didn't communicate with one another when selling tickets, so it was close to impossible to create itineraries using multiple airlines: the juggling was just too complex, even for the most experienced travel agents. It wasn't until 1987 that several airlines – Air France, Iberia, Lufthansa and SAS – created a computer system, which they named Amadeus, that allowed agents to create such complex tickets. In 2019, before airline traffic came to a halt due to the Covid-19 pandemic, Amadeus processed over six hundred million bookings – a significant chunk of all airline travel. All thanks to creating digital interfaces across airlines.

There are several other systems like Amadeus. The API economy does not create monopolies. Instead, it opens businesses up to collaborating with others: additional integrations are relatively easy. As for Amadeus, it now offers additional APIs: Air APIs, Hotel APIs, Destination Content APIs and Trip APIs. The names should speak for themselves: Amadeus is no longer just about flights.

Building a digital ecosystem is not easy, and it requires trust: a business needs to give up a lot of control over the services it offers to customers. Thankfully, providers have an incentive to deliver quality services: if they don't, it's very easy for the orchestrator to switch to a competitor. And in the economy of algorithms, it's certainly possible to fully automate such a switch.

Rule 9: Create a bold future

In 1971, a modest coffee-bean shop opened in Seattle. For the next decade, the business experienced moderate growth, operating a few stores selling coffee beans. In 1982, the business opened its fifth store, which featured a coffee bar that sold brewed coffee. Later that year a young New Yorker visited the bar and became captivated by its potential. He applied for a job with the business and became its marketing director. A year later, the same man, Howard Schultz, travelled to Italy and fell in love with coffee-house culture.[20] He envisioned introducing this culture to America, creating hubs for conversation and a sense of community. By now, you've likely guessed that the business I'm talking about is Starbucks, and you may have heard Schultz's story before – he became the company's CEO in 1986. Starbucks now operates over thirty thousand coffee shops worldwide. No matter where you are, there's a good chance you'll find a Starbucks nearby. It all began with the bold vision of one New Yorker.

In 2006, Elon Musk shared a master plan for Tesla. 'The overarching purpose of Tesla Motors,' he wrote, 'is to help expedite the move from a mine-and-burn hydrocarbon economy towards a solar electric economy.' He summarised the master plan as follows: '1. Build a sports car. 2. Use that money to build an affordable car. 3. Use *that* money to build an even more affordable car. 4. While doing above, also provide zero emission electric power generation options.'[21] Remember, Tesla was only just starting out in 2006: the first unit of its first production car, the Tesla Roadster, was built in 2008.

Ten years after sharing his initial plan for the company, Musk wrote a new one. '[1] Create stunning solar roofs with seamlessly integrated battery storage. [2] Expand the electric vehicle product line to address all major segments. [3] Develop a self-driving capability that is 10x safer than manual via massive fleet learning. [4] Enable your car to make money for you when you aren't using it.' Musk is not shy about this process – you can find his master plans on the official Tesla website.[22]

A master plan, by definition, is bold. Many business leaders are too shy to come up with something too extraordinary. They feel they don't have the freedom. I hear this sentiment quite often. In the end, there are boards, shareholders and even employees who might want to moderate the vision and make it 'safer'. Elon Musk didn't have complete freedom either. Limited technological progress and, to some degree, funding were his constraints. But he didn't shy away from aiming high. It's well over fifteen years since Musk shared his initial vision. In July 2023, the market value of Tesla was estimated at $890 billion. If you combine the market value of competing car manufacturers, such as Toyota, BMW, Volkswagen, Ford and General Motors, their total value is less than $500 billion. Not even close to Tesla's.

But perhaps I shouldn't compare Tesla to other car manufacturers. If you own a Tesla, or have driven one, you know it's more of a computer on wheels than a computerised vehicle. If you categorise Tesla as a technology company instead, it would still be among the largest of its type in the world – smaller than Apple, Alphabet (Google's mother company) and Microsoft, but bigger than Meta (formerly Facebook) or IBM. Pretty impressive for a company that didn't exist twenty years ago.

Tesla is constantly pushing boundaries and thinking ahead. Not content with simply being a car company, it instead looks for ways to innovate and provide solutions to the world's problems. It has invested heavily in the development of renewable energy, artificial intelligence and autonomous-vehicle technology. Its goal is to create a sustainable future by providing clean energy sources and smarter, safer transportation. Tesla is also investing in the research and development of new technologies, including its hyperloop project: a network of transportation tunnels built under and between gridlocked cities. Tesla not only looks to the future but is actively working to create it.

Yes, a bold vision is not worth much without a plan. But conversely, a plan can only be developed if a clear target – a vision of the future – is set. It is impossible to stay on course if you don't know where you're going. One of Jeff Bezos's mantras is to be stubborn on vision and flexible on details.[23] A plan can always be refined as you go. A strong vision helps you decide what refinements are needed.

The economy of algorithms offers businesses an opportunity unlike anything that's come before it. Scaling up is easy to do. Failing to tap into this potential for growth is the most significant mistake you could make as a business. You don't want to remain restricted to your existing markets or customers.

On 5 April 2023, Tesla released its 'Master Plan Part 3'. It's building an even bolder vision on top of its previous achievements: to revolutionise sustainable living, transportation and energy solutions for the entire planet.

Forty-five years ago, Bill Gates banged his fist on the table and asked people to pay up. At the time, almost no one thought you could make money selling software to 'the hobby market' – aka to mere mortals. Hardware, sure. But software? No way!

In the 1970s, 'hobbyists' had started to experiment with new technology: microcomputers. There was a strong culture of sharing, and clubs such as The Homebrew Computer Club were hubs for exchanging ideas and copying software. For most hobbyists, software was 'free as in free beer' and 'free as in free speech'.

Gates, who had just co-founded Micro-Soft, wasn't too enthused by this culture of sharing. He and Paul Allen had recently built their first piece of software, Altair BASIC, hiring developer Monte Davidoff to help, and Altair BASIC had become extremely popular among hobbyists. But almost none of them paid for it – why pay for something you can copy for free? – and Micro-Soft couldn't break even. So, Bill wrote and published 'An Open Letter to Hobbyists'. He pointed out that if a business puts a lot of effort into creating software, it has the right to be paid.

Even though there's logic to Gates's argument, he was seen as the 'bad guy' by hobbyists, who detested such commercial thinking. Despite a rift between those 'trying to make things happen' and those 'trying to make money', Gates achieved his goal.[24] We all know how the story ended: in less than twenty years, Gates built a de-facto software monopoly and became the wealthiest person in the world.

What made Gates special back then was not his technical skill. There were quite a few large businesses, including IBM, that were definitely better at software development. Gates's competitive advantage was his goal to create the foundations for a world in which he could thrive.

As I finish writing this book, large-scale AI algorithms are becoming increasingly capable. But it seems we are not living in a time of 'hobbyists'. To create industry-changing algorithms, huge investments are needed. GPT-3 was trained on half a trillion words. It would take a human almost three thousand years of reading[25] to read as much text. That's sixteen hours per day – day in, day out. Of course, no human had to read *any* text, because the process was fully automated. Still, due to the computing infrastructure required to create a working AI algorithm, the cost of just one such training session is several million dollars. And Sam Altman, the CEO of OpenAI, said that the cost of training GPT-4 was more than $100 million.[26] It was within OpenAI's budget: it had received a $1-billion investment from Microsoft in 2019. OpenAI transitioned from a non-profit to a for-profit company the same year. Coincidence?

When I think about it, OpenAI is doing what Microsoft did many years ago: setting the stage for a bold future in which they will thrive. Did I mention that in 2023 Microsoft committed to another multibillion-dollar investment in OpenAI?[27]

As this book was going to print, a major leadership shake-up occurred at OpenAI: Altman was fired by the company's board of directors. The reason for the termination remains unclear at this point, but there are rumours the board was uncomfortable with how fast Altman wanted to release new products, given how powerful the algorithms are becoming. Within days, Microsoft CEO Satya Nadella announced that Microsoft had hired Altman, along with OpenAI co-founder Greg Brockman, to lead their new AI team. More than 730 OpenAI employees signed an open letter threatening to leave the company in response to the decision. One day after Nadella's announcement, Altman was reinstated as OpenAI's CEO and Brockman returned to the company, under a reconstructed board of directors. These four days of drama demonstrated how quickly change can happen. When you read this book, the AI landscape might look very different. These companies are RACERS indeed.

Conclusion:
Human Agency

As we reach the conclusion of this book, it is essential to remind ourselves of the transformative power of the economy of algorithms. We are living in a world where numerous algorithms are surpassing the human ability to perform certain tasks. The rapidity with which this has occurred might feel overwhelming to many of us, leading us to question not only the security of our jobs but also our role in society.

Algorithms have been subtly supporting humanity since the earliest days of ancient Greece and steadily expanding their impact ever since. The advent of computers provided a platform for increasing their speed, previously limited by the speed of the human brain or machines such as the Analytical Engine and Jacquard loom. The emergence of the internet allowed them to significantly increase their reach, previously constrained to individual machines or small networks of computers. The recent developments in artificial intelligence have given algorithms the capacity to develop and evolve; previously, they were restricted by what humans could devise and capture in computer code. When we recognised the potential of modern algorithms to revolutionise the business world, we directed more resources towards them, and this gave rise to the economy of algorithms.

The novelty of the economy of algorithms lies in its potential to revolutionise various aspects of our lives, including how we work, communicate, make decisions and solve problems. This is reshaping our society, and if left unchecked it could lead to some people benefiting immensely while others are left behind.

However, the economy of algorithms also holds the potential to bridge divisions, such as those that exist between different industries, cultures and even nations. By enabling seamless collaboration and knowledge sharing, algorithms can contribute to a more inclusive and interconnected world, where opportunities and resources are more equitably distributed.

In developing countries, algorithms have the potential to revolutionise access to education. Large language models such as GPT-4 could serve as personal tutors: they already do a pretty good job of explaining complex concepts, and it is likely they will continue to improve. Digital minions such as the one that tried to alert my wife when I fell during a trail run could bring health-monitoring services to the most remote communities. And agricultural algorithms could offer precise farming recommendations, helping small-scale farmers optimise crop yields and manage resources efficiently, increasing their income potential.

As we have seen, algorithms have already changed our lives and our work in some beneficial ways, and we are just beginning to understand how surprisingly creative they can be. It's tempting to invite them into even more aspects of our lives and businesses. While I encourage every business I work with to consider the transformative potential of digital minions, it's important to do this responsibly. Algorithms are increasingly sophisticated, but they are not as autonomous as one might initially assume. They still need human support and supervision; in fact, the role of people is more crucial than ever. It is our responsibility to ensure these algorithms contribute to a better world, not a dystopian future.

For some, this dystopian future may already be here – especially if their work is dictated by algorithms, they are managed by algorithms or their lives are significantly impacted by algorithmic decisions. It is the collective responsibility of society to ensure that such situations do not spiral out of control.

For others, a utopian future might be unfolding. Perhaps OriHime-D pilots have a thing or two to tell us about the empowering potential of algorithms and robots. It is a complex landscape.

In order to navigate it, we must all take the time to understand the algorithms we interact with and learn how to work with them effectively. My thinking about algorithms has evolved significantly as I've researched this topic extensively over the last several years, and I hope this book provides some valuable insights to guide you on a similar journey of your own.

I opened this book with the words 'Fuck the algorithm!' – the frustrated chant of UK high-school students struggling to come to terms with the impact algorithms were beginning to have on their lives. It's an interesting choice of words. It implies that the students thought of the algorithm as an entity with agency – a trait we typically attribute to humans. As someone with a background in software development, it was initially challenging for me to comprehend this perspective. In my eyes, it was not the algorithm that was deciding the fate of the students: the algorithm was simply following instructions created by humans.

I was wrong. While the algorithm was indeed following human instructions, the way it implemented them had unforeseen consequences, which many, including the algorithm's creators, perceived as unjust. The algorithm had reached a level of autonomy that allowed it to make decisions that its creators had not intended. This behaviour can be perceived as agency. Of course, this agency is derived from their programming and the data they have been trained on, not from any inherent consciousness or desire. But as I have argued in this book, algorithms are becoming de-facto agents, albeit imperfect ones.

However, it has become clear to me that the impact of the economy of algorithms – and whether that impact will be positive or negative – does not just come down to the agency we attribute to algorithms. It's when we relinquish our own agency, failing to actively shape and oversee this new economy, that opportunities are missed and problems arise. When we step in and take control, we can drive positive outcomes and shape a better future.

In the end, it is about humans after all.

Reflecting more on the opening words of this book, I would like to offer an alternative statement that encapsulates my hopefulness about the economy of algorithms and emphasises the importance of humans to its success. I don't expect any students to shout these words in front of 20 Great Smith Street, London – they're not catchy enough – but I hope they convey what's truly important right now: Let us not surrender to their growing dominance but instead embrace our human capacity to ensure a harmonious and collaborative relationship with the algorithms that are transforming our world. Our future depends on this engagement.

Don't fuck the algorithm.

Notes

Introduction

[1] I don't know of anyone who owns a driverless vehicle. Organisations, yes, but not individuals – not yet. By 'self-driving', I mean vehicles that have the ability to navigate roads by themselves, with the constant supervision of a driver who is able to take control immediately, at any time.

[2] A.C. Clarke, 'Hazards of Prophecy: The Failure of Imagination', in A.C. Clarke, *Profiles of the Future: An Enquiry into the Limits of the Possible*, London: Victor Gollancz, 1973, p. 14.

Chapter 1

[1] Strictly speaking, all of these algorithms' responses are guesses. These particular algorithms are generating content using probabilistic methods: they try to guess the most probable outputs.

[2] Alan Turing, 'Computing Machinery and Intelligence', *Mind*, vol. LIX, no. 236 (October 1950): 433–60.

[3] C. Mauran, 'HustleGPT is a Hilarious and Scary AI Experiment in Capitalism', *Mashable*, 21 March 2023, mashable.com/article/gpt-4-hustlegpt-ai-blueprint-money-making-scheme.

[4] Because the majority of the examples I discuss in this book involve multinationals or US-based businesses, all sums cited are in US dollars unless otherwise specified.

[5] @jtmuller5, 'The-HustleGPT-Challenge', GitHub, accessed January 2023, github.com/jtmuller5/The-HustleGPT-Challenge.

[6] '275 HustleGPT Startups!', Notion, accessed 16 July 2023, thestudio1.notion.site/275-HustleGPT-Startups-1737b59d19b745ceacb8aeefe8462737.

[7] Karan Girotra, Lennart Meincke, Christian Terwiesch and Karl T. Ulrich, 'Ideas Are Dimes a Dozen: Large Language Models for Idea Generation in Innovation', working paper, 10 July 2023), ssrn.com/abstract=4526071 or dx.doi.org/10.2139/ssrn.4526071.

[8] A more formal definition of an algorithm is 'an unambiguous specification of how to solve a class of problems' or 'a self-contained step-by-step set of operations to be performed': 'Algorithm (Disambiguation)', Wikipedia, accessed 16 July 2023, en.wikipedia.org/wiki/Algorithm_(disambiguation).

9   As of October 2018, the largest known prime number is $2^{77232917} - 1$, a number with 23,249,425 digits. Just checking whether this is a prime number took six days of non-stop computing on a contemporary home computer: 'GIMPS Project Discovers Largest Known Prime Number: $2^{77232917} - 1$', Great Internet Mersenne Prime Search, 3 January 2018, mersenne.org/primes/press/M77232917.html.

10 The poor performance of the algorithm was aggravated by human decisions: staff were pressured to tinker with the algorithm and common sense was disregarded when employees questioned the algorithm's decisions: Will Parker, 'What Went Wrong with Zillow? A Real Estate Algorithm Derailed Its Big Bet', *The Wall Street Journal*, 17 November 2021.

11 I am only half joking here. Creating systems that can tell real content from fake is an important challenge for algorithm creators.

12 I prefer to think of them as *smart*.

13 J.K. Winkler, C. Fink, F. Toberer, A. Enk, T. Deinlein, R. Hofmann-Wellenhof and H.A. Haenssle, 'Association Between Surgical Skin Markings in Dermoscopic Images and Diagnostic Performance of a Deep Learning Convolutional Neural Network for Melanoma Recognition', *JAMA Dermatology*, vol. 155, no. 10, (2019): 1135–41.

14 J. Larson, S. Mattu, L. Kirchner and J. Angwin, 'How We Analyzed the COMPAS Recidivism Algorithm', *ProPublica*, 29 February 2020, propublica.org/article/how-we-analyzed-the-compas-recidivism-algorithm.

15 Academics use two terms – augmentation and automation – to describe the impact that algorithms have on work. Augmentation refers to the use of algorithms to support humans in their work. Automation refers to the replacement of humans by algorithms. In this book, I use the term automation to refer to task automation, not the automation of humans. As such, I have no need to use the term augmentation. Also, I don't like it.

16 E. Globytė, 'NordVPN Survey Shows: Cross-Device Tracking Is Widespread', NordVPN, 22 March 2023, nordvpn.com/pl/blog/spying-gadgets-research.

17 C. Stoll, 'Why the Web Won't Be Nirvana', *Newsweek*, 26 February 1995.

18 'The Spy Who Hacked Me', *Infosecurity Magazine*, 31 October 2011.

19 Remember that funny holiday photo you took with your phone last year? It would take about an hour to send at that speed.

20 Using a connection of that speed, you could send the photo in a few seconds, and you could even watch a Netflix movie, but not if you have a fancy 4K TV.

21 I take inspiration from Erik Brynjolfsson, who described how Daedalus, an Athenian craftsman in Greek mythology, went about automating the daily lives of Greeks: 'The Turing Trap – The Promise and Peril of Human-Like Artificial Intelligence', *Daedalus*, vol. 151, no. 2 (Spring 2022): 272–87.

22 As quoted by Luigi Menabrea, *Sketch of the Analytical Engine Invented by Charles Babbage Esq.*, London: Richard and John E. Taylor, 1842, p. 694.

23 A. Agostinelli, T.I. Denk, Z. Borsos, J. Engel, M. Verzetti, A. Caillon, Q. Huang, A. Jansen, A. Roberts, M. Tagliasacchi, M. Sharifi, N. Zeghidour

and C. Frank, 'MusicLM: Generating Music from Text', ArXiv, 26 January 2023, arxiv.org/abs/2301.11325.

24 Visual Analytics Research Team, 'Keynote Andy van Dam Keynote (ACMHT 2019)' (video), YouTube, 19 September 2019, youtube.com/watch?v=g0yx-F1FGnc&feature=youtu.be.

25 Yes, I know, I could just google it.

Chapter 2

1 V. Irwin, 'Elon Musk Offered a College Freshman $5k to Delete a Twitter Account', *Protocol*, 31 January 2022, protocol.com/elon-musk-flight-tracker.

2 The bot account, not the plane.

3 Paulina Cachero, 'The Teen Who Tracked Elon Musk's Jet Is Now Starting a Business to Monitor the Flights of Other Billionaires', *Bloomberg*, 28 February 2022.

4 G. Kay, T. Rains, S. Bhaimiya, H. Tan, and B. Nolan, 'Twitter Took Down the Accounts that Track Elon Musk's and Mark Zuckerberg's Jets', *Business Insider*, 15 December 2022.

5 W. Daniel, 'Elon Musk's $44 Billion Twitter Takeover Is One of the Biggest Tech Acquisitions of All Time. Here's Where the Billionaire's Deal Ranks in Tech History', Yahoo! Finance, 26 April 2022, finance.yahoo.com/news/elon-musk-44-billion-twitter-215953520.html.

6 T. Santora, 'Why Do We Blink?', Livescience.com, 6 July 2021, livescience.com/why-do-we-blink.html.

7 FidoNet still exists, but it is used only by nostalgia-driven hobbyists.

8 I owe them a lot. I also mean this in the literal sense: they never asked why our phone bills were so high.

9 N.N. Taleb, *The Black Swan: The Impact of the Highly Improbable*, Vol. 2, New York: Random House, 2007.

10 Brian Appleyard, 'Books That Helped to Change the World', *The Sunday Times*, 19 July 2009.

11 I know. I had so many questions too. Can birds use credit cards? Do turkeys have opposing thumbs, allowing them to pick up a phone?

12 The comparison might be too accurate: just as a turkey has the chance to receive a 'pardon' from the US president once a year, failing businesses – especially banks and car manufacturers – get an occasional pardon in the form of a bailout.

13 We learned a lot of such buzzwords.

14 W.C. Kim, 'Blue Ocean Strategy: From Theory to Practice', *California Management Review*, vol. 47, no. 3 (2005): 105–21.

15 Another buzzword. This section is saturated with them.

16 DARPA, or the Defense Advanced Research Projects Agency, is a research and development agency of the US Department of Defense.

17 Sebastian Thrun, 'CFree Online Class on Artificial Intelligence' (video), YouTube, 14 August 2011, youtube.com/watch?v=H9ngd6zCeUc.

18 'Sebastian Thrun', Stanford University website, accessed 16 July 2023, robots.stanford.edu/personal.html.

[19] Between 1981 and 1984, he was a chief scientist at Atari, a company that influenced my passion for computers. I still have my first Atari 800XL, as well as my first portable computer, an Atari Portfolio.

[20] N. Carlson, 'Uber Is Planning for a World Without Drivers – Just A Self-Driving Fleet', *Business Insider*, 29 May 2014.

[21] But drivers don't care anyway.

[22] Paranoid androids?

[23] Is 'slamming' the breaks a uniquely human act? The car definitely stopped abruptly.

[24] Toby Walsh, *Machines Behaving Badly: The Morality of AI*. Collingwood: Black Inc., 2022.

[25] David Shepardson, 'Backup Driver in 2018 Uber Self-Driving Crash Pleads Guilty', *Reuters*, 29 July 2023.

[26] M. Herger, 'What I've Learned After 26 Rides in a Driverless Cruise Robotaxi', The Last Driver License Holder website, 6 November 2022, thelastdriverlicenseholder.com/2022/11/05/what-ive-learned-after-26-rides-in-a-driverless-cruise-robotaxi/

[27] 'CPUC Approves Permits for Cruise and Waymo To Charge Fares for Passenger Service in San Francisco', California Public Utilities Commission, website, 10 August 2023, cpuc.ca.gov/news-and-updates/all-news/cpuc-approves-permits-for-cruise-and-waymo-to-charge-fares-for-passenger-service-in-sf-2023.

[28] Duh.

[29] M. Wayland and L. Kolodny, 'Self-Driving Cars from GM's Cruise Block San Francisco Intersection in Latest Problem for Autonomous Vehicles', CNBC, 1 July 2022.

[30] K. Wiggers, 'Cruise Acquires Driverless Vehicle Startup Voyage to Tackle Dense Urban Environments', *VentureBeat*, 15 March 2021, venturebeat.com/ business/cruise-acquires-driverless-vehicle-startup-voyage-for-an-undisclosed-amount.

[31] S. Crowe, 'Watch a Cruise Robotaxi Get Pulled Over by SF Police', *The Robot Report*, 11 April 2022, therobotreport.com/cruise-robotaxi-get-pulled-over-by-sf-police.

[32] J. Koebler, 'Food Delivery Robot Casually Drives Under Police Tape, Through Active Crime Scene', *Vice*, 16 September 2022, vice.com/en/article/93adae/food-delivery-robot-casually-drives-under-police-tape-through-active-crime-scene.

[33] S. Mellor, 'Amazon's Warehouse Problems? It's Running Out of Workers to Hire, and Has Too Much Space', *Fortune*, 21 June 2022.

[34] I do see the irony of these words: in 2018 Amazon launched Amazon Go, a chain of convenience stores without cashiers. But in fact, this just proves the point: there are only a few dozen such stores around the world – exactly twenty-seven at the time of writing this book. In just a few months at the beginning of 2023, it laid off twenty-seven thousand employees. It's a large number, but it seems like a tiny correction if you consider that in 2020 and 2021 they hired several hundred thousand new employees.

Chapter 3

[1] Or electricity.

[2] In Europe, every new model approved for manufacturing after 2018 must be equipped with a system that automatically alerts emergency services in the case of an accident. The system, called eCall, works in all EU countries and in all cars, regardless of where they were bought or registered: 'eCall 112-Based Emergency Assistance From Your Vehicle', Your Europe, 6 May 2022, europa.eu/youreurope/citizens/travel/security-and-emergencies/emergency-assistance-vehicles-ecall/index_en.htm.

[3] Smart dishwashers and washing machines were introduced around 2016. Since then, not many new models have been released. It is possible that customers are not as passionate about washing powder as I am about coffee beans: Kif Leswing, 'This Smart Washing Machine Will Order More Detergent From Amazon', *Fortune*, 5 January 2016; Sarah Perez, 'Amazon Announces New Dash-Powered Devices That Can Auto-Reorder Your Coffee, Air Filters and More', *TechCrunch*, 23 November 2016, techcrunch.com/2016/11/22/amazon-announces-new-dash-powered-devices-that-can-auto-reorder-your-coffee-air-filters-and-more.

[4] A Scrooge fridge.

[5] A hipster freezer. Is it okay if I feel bad that I have to explain these words?

[6] 'I bought them before they were cool' suddenly becomes very literal.

[7] 'Product Support', Samsung, accessed 26 July 2023, samsung.com/au/support/home-appliances/family-hub-apps.

[8] This step would become redundant once the coffee machine made all the purchases.

[9] Triggered by their concern about 'gatekeeper' smart-device manufacturers that prioritise their own products and services, the European Commission introduced the *Digital Markets Act* in 2022 to ensure such practices are weeded out.

[10] You could imagine an equivalent, subscription-based approach in which the purchasing decisions are 'outsourced'. Being a control freak, I want my coffee machine to make these decisions, not another business.

[11] This is an interesting dilemma: should it be mandatory for chatbots to disclose that they aren't human? Many believe it is unethical for an algorithm to hide its true nature.

[12] James Vincent, 'Twitter Taught Microsoft's AI Chatbot to Be a Racist Asshole in Less Than a Day', *The Verge*, 24 May 2016.

[13] Peter Lee, 'Learning From Tay's Introduction', Microsoft, blogs.microsoft.com/blog/2016/03/25/learning-tays-introduction.

[14] Kevin Roose, 'Bing's AI Chat: "I Want to Be Alive"', *The New York Times*, 16 February 2023.

[15] 'The New Bing & Edge – Updates to Chat', Microsoft Bing Blogs, accessed 16 July 2023, blogs.bing.com/search/february-2023/The-new-Bing-Edge-%E2%80%93-Updates-to-Chat.

[16] Quantopian dropped from the list the year after: 'Quantopian', *Forbes*, accessed 16 July 2023, forbes.com/companies/quantopian/?sh=371341911364.

17 Codecademy.com estimates it takes twenty-five hours to complete their beginner course – Learn Python 3 – online.

18 While Quantopian is now non-existent, many of the code examples are still available. Here is a tutorial showing a simple algorithm continually investing in Apple stocks: 'AT / Hello World Example.ph', GitHub, accessed 16 July 2023, github.com/gjhernandezp/AT/blob/master/Hello%20World%20 Example.py.

19 A. Hufford, 'Algorithmic Trading: The Play-at-Home Version', *The Wall Street Journal*, 10 August 2015.

20 Nick Bostrom, 'Ethical Issues in Advanced Artificial Intelligence', Nick Bostrom.com, accessed 16 July 2023, nickbostrom.com/ethics/ai.html.

21 Hufford, 'Algorithmic Trading'.

22 J. Kasperkevic, 'Swiss Police Release Robot That Bought Ecstasy Online', *The Guardian*, 21 February 2017.

23 The Random Darknet Shopper amassed an impressive stash of products, some of them perfectly legal, some others that … well, it got arrested for a reason. You can view the list here: wwwwwwwwwwwwwwwwwwwwww. bitnik.org/r. (Yes, that's 'w' twenty-two times, in case you're typing this link into your browser.)

24 SAE International has developed a classification system, with six levels of vehicle automation, from Level 0, 'No Driving Automation', to Level 5, 'Full Driving Automation': 'Taxonomy and Definitions for Terms Related to Driving Automation Systems for On-Road Motor Vehicles', SAE International, 30 April 2021, sae.org/standards/content/j3016_202104.

25 You might think Tesla is one of them, but they're not – at least not yet. Mercedes was the first car company to achieve level-3 certification for one of its vehicles.

26 'Tesla Driver Sleeps in Autopilot Mode', Polizeipräsidium Oberfranken, 29 December 2022, polizei.bayern.de/aktuelles/pressemitteilungen/041271/ index.html.

27 That is, stupid.

28 Tell me how old you are without telling me how old you are. If you didn't watch *The Jetsons* when you were young, try watching the cartoon now. It's still just as entertaining!

29 Robotic process automation is often perceived as focusing more on the automation of human tasks, whereas business process automation focuses on the orchestration. But in most cases, you can treat these two as synonyms.

30 Rob Nicholls, 'Stuff-Up or Conspiracy? Whistleblowers Claim Facebook Deliberately Let Important Non-news Pages Go Down in News Blackout', Pro Bono Australia, 11 May 2022, probonoaustralia.com.au/news/2022/05/ was-facebook-charity-black-out-deliberate.

31 Josh Taylor, 'Facebook Whistleblowers Allege Government and Emergency Services Hit by Australia News Ban Was a Deliberate Tactic', *The Guardian*, 6 May 2022.

32 Yes, I see some potential inconsistency here. Aren't we all a bit inconsistent in what and whom we trust?

33 Some analysts claim values are closer to 90 per cent.

34 D. Kestenbaum, '"We Built a Robot That Types": The Man Behind Computerized Stock Trading', NPR, 23 April 2015.

35 The *fastest trader in the world* made Peterffy $50 million that year. In 2021, *Forbes* magazine ranked the humble immigrant from Hungary as the sixty-fifth richest person in the world.

36 A nanosecond is one-billionth of a second.

37 Admiral Grace Hopper, a pioneer of computer programming, described 'how long a nanosecond is' in a lecture at University of Tennessee. Spoiler: it is 11.8 inches long. Her witty explanation can be viewed online: 'Admiral Grace Hopper Explains the Nanosecond' (video), YouTube, 1 March 2012, youtube.com/watch?v=9eyFDBPk4Yw.

38 Michael Eisen, 'Amazon's $23,698,655.93 Book About Flies', It Is Not Junk (blog), 22 April 2011, www.michaeleisen.org/blog/?p=358.

39 This sounds like a mistake in the bot's code. Perhaps the intention was to increase the price by 0.27 per cent, not 27 per cent. On the other hand, perhaps bordeebook didn't have the book at all and the extra margin was meant to cover the costs of buying the book from another seller – presumably profnath – first!

Chapter 4

1 S. Evans, 'Coronavirus Australia: Bonds Owner Rushing Four Million Masks Down Under', *The Australian Financial Review*, 3 August 2020.

2 About two years later, I used the same strategy to buy a car. The model was about to launch in Australia, and my wife's car needed to be replaced. The car was in high demand, a lot of people wanted to buy this particular model at the time. I bought it about fifteen minutes after it was added to the website – this particular manufacturer sells cars mostly online. A friend of mine, who was also planning to buy the same car, ordered it an hour later, and took the delivery a month later. This was the difference between hiring an algorithm to buy the car and buying it 'manually'.

3 M. Weber, M. Kowalkiewicz, J. Weking, M. Bohm and H. Krcmar, 'When Algorithms Go Shopping: Analyzing Business Models for Highly Autonomous Consumer Buying Agents', Proceedings of the Fifteenth International Conference on Wirtschaftsinformatik, Potsdam, Germany, March 2020, pp. 1–16.

4 Perhaps I should mention that my co-authors and I came up with the name 'highly autonomous consumer buying agents' in a research paper.

5 Strictly speaking, they don't *know* what we want, but they are predicting our needs, and often these predictions are quite accurate, making us feel like they *know*.

6 European Commission, 'Antitrust: Commission Opens Investigation Into Possible Anti-competitive Conduct of Amazon', Press Release, 17 July 2019, ec.europa.eu/commission/presscorner/detail/en/ip_19_4291.

7 Cheesy? I couldn't help myself.

8   T. Ackerman, 'A Clayton County Man Is Still Alive Thanks to His Apple Watch and a 911 Dispatcher', *MSN*, 5 February 2022.

9   It's become quite obvious that one of the points I am making in this book is that we must prepare our systems to interact with new types of agents. These new agents often behave like humans – they can speak and make phone calls – but they also behave unlike humans (in all sorts of quirky ways) – they might use degrees, minutes and seconds to describe a location.

10 Is the algorithm sharing its insights with anyone else?

11 Yes, it reminds me to buy filters, but not coffee beans. How annoying!

12 T. Jones, 'How HP's Instant Ink Will Work in Australia', Gizmodo Australia, 15 April 2021, gizmodo.com.au/2021/04/how-hps-instant-ink-subscription-will-work-in-australia-price.

13 P. Kunert, 'HP Pilots Paper Delivery Service for Instant Ink Subscribers', *The Register*, 9 June 2022.

14 Kunert, 'HP Pilots Paper Delivery Service'.

15 J. Davidson, B. Liebald, J. Liu, P. Nandy, T. Van Vleet, U. Gargi, S. Gupta, Y. He, M. Lambert, B. Livingston and D. Sampath. 'The YouTube Video Recommendation System', Proceedings of the Fourth ACM Conference on Recommender Systems, September 2010, pp. 293–96.

16 X. Amatriain and J. Basilico. 'Netflix Recommendations: Beyond the Five Stars', *Medium*, 6 April 2012.

17 S. Liberatore, 'Sony PlayStation 5 Sale Causes Walmart Website to Crash', *Daily Mail*, 12 November 2020.

18 A. Cranz, 'The Needless Drama of Buying a PS5', *The Verge*, 23 May 2021, theverge.com/2021/5/23/22434905/ps5-restock-best-buy-walmart-sony-playstation-5.

19 Gamers play and scientists work, but also the other way round! E-sports players earn serious money playing games, and many data scientists have the best of fun while crunching numbers.

20 'Are Bots Ruining the Yosemite Reservation System?', Tripadvisor, accessed 16 July 2023, tripadvisor.com.au/ShowTopic-g61000-i315-k11364005-o10-Are_Bots_Ruining_the_Yosemite_Reservation_System-Yosemite_National_Park_California.html.

21 I expect this to change, as tools like ChatGPT can act as personal software developers and build simple bots that customers can run themselves.

22 H. Nagarajan, 'Hari-Nagarajan/Fairgame', GitHub, 2021, github.com/Hari-Nagarajan/fairgame.

23 Weber et al., 'When Algorithms Go Shopping'.

24 You might prefer to call them scalpers, depending on how you view the morality of their actions.

25 M. Kan, 'How Do Bots Buy Up Graphics Cards? We Rented One to Find Out', *PC Magazine Australia*, 21 April 2021.

26 @Jbrowder1, 'DoNotPay Will Pay Any Lawyer or Person ...', Twitter, 9 January 2023, twitter.com/jbrowder1/status/1612312707398795264.

27 @Jbrowder1, 'Good Morning! Bad News', Twitter, 26 January 2023, twitter.com/jbrowder1/status/1618265395986857984.

28 S. Seddon, 'You Can Now Buy Virgin Train Tickets on Your Amazon Alexa Device', *Chronicle Live*, 3 May 2018.

29 Calling your digital minion 'mate' is optional.

Chapter 5

1 @FiletOFish1066, 'Finally Fired After Six Years', Reddit, 23 May 2016, web.archive.org/web/20160523114950/https://www.reddit.com/r/cscareerquestions/comments/4km3yc/finally_fired_after_6_years.

2 The original Reddit post was deleted but can be accessed thanks to Wayback Machine: @FiletOFish1066, 'Finally Fired After Six Years'. Leo Tiffish also deleted his Reddit account.

3 I meet a lot of professionals who complain they are too busy to think about optimising their work (which means finding ways to be less busy). They're also too busy to fully comprehend the virtuous cycle they are in.

4 Karel Čapek, *R.U.R.*, translated by Paul Selver and Nigel Playfair, New York: Dover Publications, 2001.

5 Indeed, I've found that an easy way to get roboticist to engage in a fiery argument is to ask them to define what a robot is – the very focus of their research.

6 J.H. Richardson, 'AI Chatbots Try to Schedule Meetings – Without Enraging Us', *Wired*, 24 May 2018.

7 X.ai was acquired in 2021 by Bizzabo, an event-planning business. Amy and Andrew retired on 31 October 2021. The domain name was purchased by Elon Musk and now hosts xAI, a cryptic organisation, with a goal to 'understand the true nature of the universe'. There are numerous alternatives to x.ai, but so far, none has managed to replicate its conversational style. It is also unclear how much of Amy's behaviour was 'real' – there are multiple reports that x.ai hired humans to take over tasks that were too hard for the algorithm. In a twist of a twist, humans were allegedly hired to pretend to be a bot. With the advent of large language models, we should see more of such algorithms emerge soon!

8 A. Sweigart, *Automate the Boring Stuff With Python: Practical Programming for Total Beginners*, San Francisco: No Starch Press, 2019.

9 @Rushmid, 'I Just Automated 100 hrs of Labor PER Week at My Company', Reddit, 23 October 2015, reddit.com/r/Automate/comments/3pu5kd/i_just_automated_100_hrs_of_labor_per_week_at_my.

10 Go ahead and define a shortcut like this right now – most operating systems, including your phone, will have an 'autocomplete' section. If you don't know where to start, enter '/thx' as a shortcut for 'Thank you for your email!' You're now saving five seconds every time you type '/thx'.

11 @Oasis1272, 'Should I Tell My Managers About a Computer Program I Wrote That Saves Me Hours of Work Every Day?' Reddit, 22 December 2016, reddit.com/r/legaladvice/comments/5jp6zs/should_i_tell_my_managers_about_a_computer.

12 @Oasis1272, '(Update) I Wrote a Computer Program On My Work Computer . . .', Reddit, 2016, reddit.com/r/legaladvice/comments/3xblof/update_i_wrote_a_computer_program_on_my_work.

13 @CS-NL, '[UPDATE] My Friends Call Me a Scumbag . . .', Reddit, 27 June 2012, reddit.com/r/AskReddit/comments/vomtn/update_my_friends_call_me_a_scumbag_because_i.

14 'Facebook Extends Its Work-At-Home Policy to Most Employees (cnbc.com)', *Hacker News*, 9 June 2021, news.ycombinator.com/item?id=27453909: see the comments section.

15 @Dreyfan didn't disclose their gender.

16 K. Darling, 'Children Beating Up Robot Inspires New Escape Maneuver System', *IEEE Spectrum*, 6 August 2015, spectrum.ieee.org/children-beating-up-robot.

17 D. Brščić, H. Kidokoro, Y. Suehiro and T. Kanda, 'Escaping From Children's Abuse of Social Robots', Proceedings of the Tenth Annual ACM/IEEE International Conference on Human–Robot Interaction, March 2015, pp. 59–66. See also: T. Nomura, T. Uratani, T. Kanda, K. Matsumoto, H. Kidokoro, Y. Suehiro, and S. Yamada, 'Why Do Children Abuse Robots?', Proceedings of the Tenth Annual ACM/IEEE International Conference on Human–Robot Interaction Extended Abstracts, March 2015, pp. 63–64.

18 M. Keijsers, H. Kazmi, F. Eyssel and C. Bartneck, 'Teaching Robots a Lesson: Determinants of Robot Punishment', *International Journal of Social Robotics*, vol. 13, no. 1 (2021): 41–54.

19 P. Liu, S. Zhai and T. Li, 'Is It OK to Bully Automated Cars?', *Accident Analysis and Prevention*, vol. 173 (August 2022): article no. 106714.

20 C. Bartneck and M. Keijsers, 'The Morality of Abusing a Robot', *Paladyn, Journal of Behavioral Robotics*, vol. 11, no. 1, (2020): 271–83.

21 R. Sparrow, 'Kicking a Robot Dog', 2016 Eleventh ACM/IEEE International Conference on Human–Robot Interaction (HRI), March 2016, p. 229.

22 The Department of Human Services was renamed, becoming Services Australia, in 2019.

23 T. Rinta-Kahila, I. Someh, N. Gillespie, M. Indulska and S. Gregor, 'Algorithmic Decision-Making and System Destructiveness: A Case of Automatic Debt Recovery', *European Journal of Information Systems*, vol. 31, no. 3 (2021): 313–38.

24 *Royal Commission into the Robodebt Scheme Report*, Canberra: Commonwealth of Australia, 2023.

25 'Automated Decision Making and Administrative Law – a Nationwide Conversation on Law Reform', Australian Law Reform Commission, 13 September 2020, alrc.gov.au/news/automated-decision-making-and-administrative-law-a-nationwide-conversation-on-law-reform.

26 Ryuzo Suzuki, 'Japanese Avatar Robots Provide Disabled Chance to Work', *Star Advertiser* (Honolulu), 18 July 2023.

27 'OriHime-D', Asratac, 13 November 2019, asratec-co-jp.translate.goog/ portfolio_page/orihime-d/?_x_tr_sl=ja&_x_tr_tl=en&_x_tr_hl=en&_x_ tr_pto=sc.

28 K. Takeuchi, Y. Yamazaki and K. Yoshifuji, 'Avatar Work: Telework for Disabled People Unable to Go Outside by Using Avatar Robots', Companion of the 2020 ACM/IEEE International Conference on Human-Robot Interaction, March 2020, pp. 53–60.

29 'Demibots' has a nice ring to it, doesn't it?

30 M. Sainato, '"I'm Not a Robot": Amazon Workers Condemn Unsafe, Grueling Conditions at Warehouse', The Guardian, 27 July 2020.

31 S. Kessler, 'Amazon Built One of the World's Most Efficient Warehouses by Embracing Chaos', Quartz, 21 February 2018, classic.qz.com/perfect-company-2/1172282/this-company-built-one-of-the-worlds-most-efficient-warehouses-by-embracing-chaos.

32 And, in some of its regions, same-day deliveries.

33 N. Wingfield, 'Amazon Brings in the Robots, Forcing Humans to Find New Roles', The Sydney Morning Herald, 11 September 2017.

34 A robot is simply an algorithm with a capable body.

35 Kacie Kinzer, Tweenbots.com, accessed 25 July 2022, tweenbots.com.

36 Director of Algorithm Assessment and Technology Insight CMA1559, Candidate information pack Competition & Markets Authority, UK Gov.

37 K. Stowers, L.L. Brady, C.J. MacLellan, R. Wohleber and E. Salas, 'Improving Teamwork Competencies in Human–Machine Teams: Perspectives From Team Science', Frontiers in Psychology, vol. 12 (May 2021): 1669.

38 The fact that many such platforms do not consider these human drivers as their employees, and call them 'partners', is a concern worthy of another book.

39 Jeffrey Dastin 'Amazon Scraps Secret AI Recruiting Tool That Showed Bias Against Women', Reuters, 11 October 2018.

40 Rudolf Siegel, 'The Impact of Electronic Monitoring on Employees' Job Satisfaction, Stress, Performance and Counterproductive Work Behavior: A Meta-Analysis', Computers in Human Behavior Reports, vol. 8 (December 2022): 100227.

41 Shuaib Ahmed, Yasir Mansoor Kundi and Nasib Dar, 'Effects of Electronic Performance Monitoring on Employee Work Engagement: A Multilevel Investigation', Academy of Management Proceedings, vol. 2022, no. 1 (July 2022).

42 Mareike Möhlmann and Ola Henfridsson, 'What People Hate About Being Managed by Algorithms, According to a Study of Uber Drivers', Harvard Business Review, vol. 30 (2019).

43 Sainato, '"I'm not a robot"'.

44 E. O'Byrne, 'Slave to the Algorithm? What It's Really Like to Be a Deliveroo Rider', Irish Examiner, 20 February 2019.

45 S.W. Sweeney, 'Uber, Lyft Drivers Manipulate Fares at Reagan National Causing Artificial Price Surges', WJLA News, 16 May 2019.

46 'The first rule of surge club is you don't talk about surge club,' I read in an Uber driver forum. Apparently, the drivers on TV didn't get the memo.

47 A. Styles, 'Uber Frustrating: Perth Rideshare Drivers Find New Ways to Price-Gouge', *The Sydney Morning Herald*, 18 February 2022.

48 Algorithmic Justice League website, accessed 16 July 2023, www.ajl.org.

49 I'm pretty sure my take on this is clear by now, but just in case it isn't: algorithms, including artificial intelligence, do not have a conscience. It's just a figure of speech, used to avoid technical language such as 'dealing with out-of-domain cases'.

Chapter 6

1 Clayton M. Christensen, *The Innovator's Dilemma: When New Technologies Cause Great Firms to Fail*, Boston: Harvard Business School Press, 1997.

2 Tesla sold less than 2500 units of its first car, the Tesla Roadster, between 2008 and 2012. It helped them finance further growth. Any traditional car manufacturer would not even consider such a 'distraction' as selling a few hundred vehicles a year. In 2021, Toyota reported production levels of over twenty thousand cars *per day*.

3 R. Bellis, 'Why These Tech Companies Keep Running Thousands of Failed Experiments', *Fast Company*, 9 September 2016, fastcompany. com/3063846/why-these-tech-companies-keep-running-thousands-of-failed.

4 Jeff Sparrow, '"Full-on Robot Writing": The Artificial Intelligence Challenge Facing Universities', *The Guardian*, 19 November 2022.

5 Rachel Metz, 'AI Won an Art Contest, and Artists Are Furious', *CNN Business*, 3 September 2022.

6 @GenelJumalon, 'TL;DR – Someone Entered an Art Competition with an AI-Generated Piece and Won the First Prize …', Twitter, 31 August 2022, twitter.com/fluxophile/status/1564723321396150272?s=20&t=bQ0Qw4 MYd_Vl3GjuLEPGeQ.

7 HoloLens is an augmented-reality headset built by Microsoft. Imagine wearing a pair of oversized glasses and, as you look around, you see digital images, text or animations layered on top of the real environment. It could be very useful for architects, who could go to a future building site and walk around a building they're designing, even before it's built.

8 C. McGarrigle, 'Forget the Flâneur', Proceedings of the 19th International Symposium of Electronic Art, Sydney, 7–16 June 2013.

9 F. Gino, 'The Business Case for Curiosity', *Harvard Business Review*, vol. 96, no. 5 (2018): 48–57; Steven Eppinger, 'Key Business Strategies for Stimulating Curiosity – Design Thinking Meets Agile', MIT Management Executive Education, 27 February 2022, exec.mit. edu/s/blog-post/key-business-strategies-for-stimulating-curiosity-MCQSOGKIANZRDDDBZYO2V2RORXYM.

10 J. Aroles and W. Küpers, 'Flânerie as a Methodological Practice for Explorative Re-Search in Digital Worlds', *Culture and Organization*, vol. 28, no. 5 (2022): 1–14.

11 H. Bohnacker, B. Gross, J. Laub and C. Lazzeroni, *Generative Design: Visualize, Program and Create With Processing*, New York: Princeton Architectural Press, 2012.

12 It's almost like swiping right or left on a dating app.

13 It might be obvious to humans that a chair should be touching the floor instead of floating in the air. It is not obvious at all to an algorithm.

14 P. Starck, 'A.I. – Introducing the First Chair Created With Artificial Intelligence', STARCK, 24 January 2020, www.starck.com/a-i-introducing-the-first-chair-created-with-artificial-intelligence-p3801.

15 K. Schwab, 'This Is the first Commercial Chair Made Using Generative Design', *Fast Company*, 17 April 2019, fastcompany.com/90334218/this-is-the-first-commercial-product-made-using-generative-design.

16 'The A.I. Chair', Gessato, 23 November 2020, gessato.com/artificial-intelligence-chair.

17 A specific mechanism in an algorithm that helps it determine whether it is evolving in the right direction is called, aptly, a fitness function.

18 No, algorithms do not hallucinate.

19 No, algorithms do not experience feelings the way humans do. They just process inputs.

20 As you can imagine, we have developed a few strategies in the past, and the process is quite consistent.

21 As you'll remember, OpenAI is the organisation that created GPT-3. They also kindly made it available via so called APIs – application programming interfaces – so that anyone can write an algorithm that uses GPT-3.

22 R. Waters, 'Automated Company Raises Equivalent of $120M in Digital Currency', *CNBC*, 17 May 2016.

23 The formal name for it is 'distributed ledger'.

24 Ethereum refers to it as a 'distributed state machine'.

25 Ethereum calls it 'running a node'.

26 In Ethereum, the computer time is priced using a unit called 'gas'. You pay for gas using gwei (one-billionth of Ether). That's enough blockchain jargon for this chapter.

27 Stephen Tual, 'Ethereum Launches', Ethereum Foundation Blog, 30 July 2015, blog.ethereum.org/2015/07/30/ethereum-launches.

28 Vitalik Buterin, 'Bootstrapping a Decentralized Autonomous Corporation: Part I', *Bitcoin Magazine*, 19 September 2013; 'DAOs Are Not Corporations: Where Decentralization in Autonomous Organizations Matters', Vitalik Buterin's website, 20 September 2022, vitalik.ca/general/2022/09/20/daos.html.

29 Not everyone agrees that true neutrality can be achieved, thus I am suggesting that algorithms can help an organisation to be *more* neutral.

30 Mars Climate Orbiter Mishap Investigation Board, 'Mars Climate Orbiter Mishap Investigation Board Phase I Report', 10 November 1999, llis.nasa.gov/llis_lib/pdf/1009464main1_0641-mr.pdf.

31 NASA calls this activity 'angular momentum desaturation'.

32 If you're about to make the point that units are essential in these examples, wait until the next paragraph and consider applying to NASA – they need people like you.

33 Here's a more down-to-earth comparison: imagine if someone told you that a hot-springs lake had a water temperature of 100 degrees. Before jumping into the lake, it would be best to clarify whether the temperature was in Celsius or Fahrenheit.

34 National Aeronautics and Space Administration, 'Lost in Translation', *System Failure Case Studies*, vol. 3, no. 5 (August 2009).

35 National Aeronautics and Space Administration, '1998 Mars Missions', Press Kit, December 1998.

36 M. Dino, V. Zamfir and E.G. Sirer, 'A Call for a Temporary Moratorium on The DAO', Hacking Distributed, 27 May 2016, hackingdistributed. com/2016/05/27/dao-call-for-moratorium.

37 M.E. Peck, 'Ethereum's $150-Million Blockchain-Powered Fund Opens Just as Researchers Call For a Halt', *IEEE Spectrum*, 28 May 2016, spectrum. ieee.org/ethereums-150-million-dollar-dao-opens-for-business-just-as-researchers-call-for-a-moratorium.

38 K. Finley, 'A $50 Million Hack Just Showed That the DAO Was All Too Human', *Wired*, 18 June 2016.

39 Such a change is called a 'hard fork'. It is like building a parallel universe in which a terrible event you'd like to avoid does not happen. Only easier.

40 N. Patel, 'From a Meme to $47 Million: ConstitutionDAO, Crypto, and Crowdfunding', *The Verge*, 7 December 2021, theverge.com/22820563/ constitution-meme-47-million-crypto-crowdfunding-blockchain-ethereum-constitution.

41 Given how 'open' DAOs are, I am still puzzled that someone believed they could win the auction. If you know the highest amount your bidders are willing to spend, it makes your bidding easy. And anyone could find out how much funds ConstitutionDAO had.

42 K. Kudelska, 'Blockchain Company Buys Land as a Group Under Unique Wyoming Law', Wyoming Public Media, 25 January 2022, wyomingpublicmedia.org/open-spaces/2022-01-25/blockchain-company-buys-land-as-a-group-under-unique-wyoming-law.

43 S. Fitsimones, 'Could a DAO Build the Next Great City? (video), TED Talk, 26 July 2022, ted.com/talks/scott_fitsimones_could_a_dao_build_ the_next_great_city.

44 'We're Tackling Aging With the Power of a Global Community', VitaDAO, accessed 16 Jul 2023, vitadao.com.

45 Seedclub website, accessed 16 July 2023, seedclub.xyz.

46 A. Hayward, 'Basketball DAO, Dogecoin Wallet App to Buy BIG3 Team Stakes via NFTs', *Decrypt*, 4 May 2022, decrypt.co/99452/basketball-dao-dogecoin-wallet-app-big3-team-stakes-nfts.

47 Marek Kowalkiewicz, 'Artificial Dreams', *Digital Economy*, 8 March 2022, marekkowal.substack.com/p/artificial-dreams.

48 'Submarines', Zia X Aza, accessed 5 September 2022, ziaxaza.com.

[49] Roose, 'An A.I.-Generated Picture Won an Art Prize'.

Chapter 7

[1] I keep a list of words that academics love but which I would hate to introduce in this book. 'Affordances', 'augmentation' or 'task discretion', anyone? You won't find these terms in this book. Apart from that one note, earlier in the book, where I mentioned one of these terms to say I don't like it.

[2] While trying not to get stuck in the Innovator's Dilemma.

[3] Tyler Cowen, 'Garry Kasparov on AI, Chess, and the Future of Creativity', *Medium*, 10 May 2017, medium.com/conversations-with-tyler/garry-kasparov-tyler-cowen-chess-iq-ai-putin-3bf28baf4dba.

[4] Max Nisen, 'Humans Are on the Verge of Losing One of Their Last Big Advantages Over Computers', *Business Insider*, 6 November 2013, businessinsider.com/computers-beating-humans-at-advanced-chess-2013-11.

[5] 'How SME Watkins Steel Transformed from Traditional Steel Fabrication to Digital Service Provision', *MIS Quarterly Executive*, vol. 21, no. 3 (2022): 4.

[6] 'Back [then],' Des Watkins explained, 'I thought innovation was just another buzzword. [Our business] is steel fabrication. You cut steel, and you weld steel. You cannot innovate it. It is a traditional industry. But, I reckon, no matter what industry you are in, there would be a better way of doing it. So, it is just the creation and delivery of new customer value … We went from the bottom of the food chain. We were just moving our way up. Then, all of a sudden, we are running the [construction] site.': Ibid.

[7] I am sure Watkins would disagree here and say this was intentional.

[8] And a hundred times in the last eleven years!

[9] M. Zeng, 'Alibaba and the Future of Business', *Harvard Business Review*, vol. 96, no. 5 (September 2018): 88–96.

[10] Ibid.

[11] K. Roose, *Futureproof: Nine Rules for Humans in the Age of Automation*, London: Random House, 2022.

[12] 'NetDragon Appoints its First Virtual CEO', *Cision PR Newswire*, 26 August 2022, prnewswire.com/news-releases/netdragon-appoints-its-first-virtual-ceo-301613062.html. I looked up their webpage while writing this chapter, a few months later, and the current 'Rotating CEO', Liu Luyuan, is distinctively human. I guess the algorithmic CEO didn't do too well.

[13] Kevin Riera, Anne-Laure Rousseau and Clément Baudelaire, 'Doctor GPT-3: Hype or Reality?', Nabla Copiolot, 27 October 2020, nabla.com/blog/gpt-3.

[14] 'Wikipedia is 20, and Its Reputation Has Never Been Higher, *The Economist*, 9 January 2021.

[15] 'Wikipedia:Size of Wikipedia', Wikipedia, accessed 11 October 2022, en.wikipedia.org/wiki/Wikipedia:Size_of_Wikipedia.

[16] Dimi Dimitrov, 'Meet "Cluebot Ng", an AI Tool to Tackle Wikipedia Vandalism', Free Knowledge Advocacy Group EU, 27 October 2021, wikimedia.brussels/meet-cluebot-ng-an-anti-vandal-ai-bot-that-tries-to-

detect-and-revert-vandalism; 'Wikipedia:Purpose', Wikipedia, accessed 11 October 2022, en.wikipedia.org/wiki/Wikipedia:Purpose.

17 Cluebot NG is operated by Wikipedia users Crispy1989, Cobi, Rich Smith, and Damian Zaremba.

18 Dimitrov, 'Meet "Cluebot Ng"'.

19 @Elonmusk, 'Yes, Excessive Automation at Tesla Was a Mistake ...', Twitter, 14 April 2018, twitter.com/elonmusk/status/984882630947753984.

20 My favourite take on this comes from Twitter user @internetofshit: 'There are like 137271882 boring ass ways this website will break like some dell shitbox running windows server 2008 r2 blue screening and nobody was around to notice', Twitter, 18 November 2022. In other words, software needs updates, hardware can break. Algorithms need constant TLC.

21 J. Ruane and A. McAfee, 'What a DAO Can – and Can't – Do', *Harvard Business Review*, 10 May 2022.

22 H.J. Wilson and P.R. Daugherty, 'Human Machine: Reimagining Work in the Age of AI', *Harvard Business Review*, 16 August 2018.

23 A highly controversial move, given the workers were paid less than US$2 per hour: Billy Perrigo, 'Exclusive: OpenAI Used Kenyan Workers on Less Than $2 Per Hour to Make ChatGPT Less Toxic', *Time*, 18 January 2023.

24 Chloe Xiang, 'Scientists Increasingly Can't Explain How AI Works', *Vice*, 2 November 2022.

25 A. Thurai and J. McKendrick, 'Overcoming the C-Suite's Distrust of AI', *Harvard Business Review*, 23 March 2022.

26 B. Christian and T. Griffiths, *Algorithms to Live By: The Computer Science of Human Decisions*, New York: Macmillan, 2016.

27 Perhaps in future we will consider decision-making in environmental protection a 'dull' task, but right now it is still a relatively 'okay' activity for humans to perform.

28 In July 2023, Musk rebranded Twitter. It's now known as X.

29 Antonio Pequeño, 'Fidelity Values Twitter at Roughly $15 Billion – a Third of the Sticker Price Paid by Musk', *Forbes*, 30 May 2023; Joe Toppe, 'Elon Musk Says Twitter Had Many Employees but Little Product Development', *FOX Business*, 18 April 2023.

30 Stephen M. Kosslyn, 'Are You Developing Skills That Won't Be Automated', *Harvard Business Review*, 25 September 2019.

Chapter 8

1 'New Car Market Up As Plate Change September Marks One Million EV Milestone', SMMT, 5 October 2022, smmt.co.uk/2022/10/new-car-market-up-as-plate-change-september-marks-one-million-ev-milestone.

2 Robin Blades, 'AI Generates Hypotheses Human Scientists Have Not Thought Of', *Scientific American*, 28 October 2021.

3 'AI Tool Accelerates Discovery of New Materials', University of Liverpool, 21 September 2021, /news.liverpool.ac.uk/2021/09/21/new-ai-tool-developed-by-liverpool-researchers-accelerates-discovery-of-new-materials.

4   Shannon Liao, 'Amid Losses, Netflix Bets on a Bold Strategy Around Video Games', *The Washington Post*, 22 April 2022.

5   As Netflix cracks down on password sharing, these numbers are likely to change.

6   R.G. McGrath,' Transient Advantage', *Harvard Business Review*, vol. 91, no. 6 (2013): 62–70.

7   'Amazon.com Launches Web Services: Developers Can Now Incorporate Amazon.com Content and Features Into Their Own Web Sites; Extends "Welcome Mat" for Developers', Amazon, 16 July 2002, press.aboutamazon.com/news-releases/news-release-details/amazoncom-launches-web-services.

8   Benjamin Black, 'EC2 Origins', Benjamin Black Causes Trouble Here, 25 January 2009, blog.b3k.us/2009/01/25/ec2-origins.html.

9   Simon Sharwood, 'Google Cloud Makes Its First Profit, 15 Years After Launching', *The Register*, 26 April 2023.

10   J. Weking, M. Stöcker, M. Kowalkiewicz, M. Böhm and H. Krcmar, 'Leveraging Industry 4.0: A Business Model Pattern Framework, *International Journal of Production Economics*, vol. 225 (2020): 107588.

11   Marek Kowalkiewicz, 'Creativity Not Required: How Great Minds Craft Ideas' (Video), TEDx Talks, 31 May 2017, YouTube, accessed 15 September 2022, youtube.com/watch?v=iJU3KoZcUKU.

12   Queensland Government, 'Queensland Licences One Step Closer to Going Digital', Media statement, 24 February 2022, accessed 15 September 2022, statements.qld.gov.au/statements/94549.

13   Brian Christian, 'The A/B Test: Inside the Technology That's Changing the Rules of Business', *Wired*, 25 April 2012.

14   M. Hawkins, 'I Let Algorithms Randomize My Life for Two Years', TED Talks, 16 February 2021.

15   Ibid.

16   I was unable to confirm whether Hawkins went for any of the random tattoo ideas.

17   Oh, the irony.

18   Hawkins, 'I Let Algorithms Randomize My Life'.

19   J. Smialek, 'The Nobel in Economics Goes to Three Who Find Experiments in Real Life', *The New York Times*, 11 October 2021.

20   Sinan Aral and Christos Nicolaides, 'Exercise Contagion in a Global Social Network', *Nature Communications*, vol. 8 (18 April 2017): 14753.

21   The very first sentence of the abstract is, 'We leveraged exogenous variation in weather patterns across geographies to identify social contagion in exercise behaviours across a global social network.' It could just as well say, 'We used the randomness of weather to see if runners run more if their social network friends run more.' But then, it probably wouldn't have been accepted by the journal *Nature*.

22   C. Ogden, 'Google Graveyard', Killed by Google, accessed 29 September 2022, killedbygoogle.com.

23 'Google: Revenue Distribution, 2017–2021, by Segment', *Statistica*, accessed 29 September 2022, statista.com/statistics/1093781/distribution-of-googles-revenues-by-segment.

24 Even school principals skip lectures. I thought 50 per cent attendance was a safe assumption. I wasn't wrong.

25 The titles were: (1) What If I Told You That Your Children's Future Is Already Here; (2) How Is AI Changing the Way We Think About Education?; (3) What If Every Child Had a Personal Tutor?; (4) How AI Is Making Us Rethink What It Means to Be Human; (5) How AI Is Changing the Way We Interact With Our Children; (6) What If AI Could Help Your Child Get Into Their Dream School?; (7) How AI Is Changing the Way We Parent; (8) What If AI Could Help Your Child Get a Job They Love?; (9) What If AI Could Help Your Child Live a Longer, Healthier Life?; (10) The Future of AI and Our Children: A Bright Future Ahead!

26 Arthur C. Clarke, 'Hazards of Prophecy: The Failure of Imagination', *Profiles of the Future: An Enquiry into the Limits of the Possible*, New York: Harper & Row, 1973, pp. 14, 21, 36.

27 D. Pathak, P. Agrawal, A.A. Efros and T. Darrell, 'Curiosity-Driven Exploration by Self-Supervised Prediction', *Proceedings of the 34th International Conference on Machine Learning*, vol. 70 (July 2017): 2778–87.

28 'The Future Enterprise', Queensland University of Technology, 12 September 2022, qut.edu.au/engage/the-future-enterprise-webinar-series.

29 Pathak et al., 'Curiosity-Driven Exploration'.

30 The main character of the Doom video game is called … the Doom Guy: 'Doomguy', Wikipedia, 9 July 2023, en.wikipedia.org/wiki/Doomguy.

31 Strictly speaking, it was 'getting rid of' its curiosity by answering the unknowns.

Chapter 9

1 L. Mirani, 'The Age of $25 Smartphones Is Upon Us – and Mark Zuckerberg Wants to Give Them a Dial Tone', *Quartz*, 24 February 2014, qz.com/180497/the-age-of-25-smartphones-is-upon-us-and-mark-zuckerberg-wants-to-give-them-a-dial-tone.

2 They might go offline for a bit when they're being updated.

3 M. Weiser, 'The Computer for the 21st Century', *Scientific American*, September 1991, pp. 94–104.

4 Also called 'pervasive computing' by those – like me – who can never spell 'ubiquitous' properly.

5 The only alternative to using algorithms would be to send company employees to constantly accompany their customers. I am sure you'll agree this would be creepy. Somehow, we're more forgiving of algorithms constantly following us.

6 J. Bensley, '2016 GBC: Where Are They Now? Smart Toilet Company', QUTeX Blog, 13 July 2021, blogs.qut.edu.au/qutex/2019/04/04/2016-gbc-where-are-they-now-smart-toilet-company.

7 Ibid.
8 Andrew Evers and Erin Black, 'This $7000 Smart Toilet Has Built-in Speakers, Mood Lighting and Amazon Alexa Voice Controls', *CNBC*, 8 January 2019.
9 This takes the need for instant deliveries to a completely new level.
10 I think it needs a better name. How about 'TrackAPoo'?
11 R. Raphael, 'Netflix CEO Reed Hastings: Sleep Is Our Competition', *Fast Company*, 6 November 2017, fastcompany.com/40491939/netflix-ceo-reed-hastings-sleep-is-our-competition.
12 'Netflix's Founder on Building an Iconic Company', Kleiner Perkins, 16 September 2015, .kleinerperkins.com/perspectives/netflix-the-new-establishment-reed-hastings-on-building-a-great-company.
13 This translated into an explicit listing of wine as a competitor on their website, which I cited in the previous chapter.
14 C.M. Christensen, T. Hall, K. Dillon, and D.S. Duncan, 'Know Your Customers' Jobs to Be Done', *Harvard Business Review*, vol. 94, no. 9 (2016): 54–62.
15 While the 'faster horse' quote is often attributed to Henry Ford, no reliable source has been identified. Regardless, the point holds: 'Henry Ford, Innovation, and That "Faster Horse" Quote', *Harvard Business Review*, 23 July 2014, hbr.org/2011/08/henry-ford-never-said-the-fast.
16 K. Westerman, A. Reaver, C. Roy, M. Ploch, E. Sharoni, B. Nogal and G. Blander, 'Longitudinal Analysis of Biomarker Data from a Personalized Nutrition Platform in Healthy Subjects', *Scientific Reports*, vol. 8, no. 1 (2018): 1–10.
17 We can be quite certain about this one – Henry Ford, in his autobiography, admitted saying it in 1909.
18 'InsideTracker Raises $15M to Drive Continued Innovation and Delivery of Its Leading AI Platform for Personalized Nutrition and Healthspan Optimization', *Cision PR Newswire*, 8 September 2022, prnewswire.com/news-releases/insidetracker-raises-15m-to-drive-continued-innovation-and-delivery-of-its-leading-ai-platform-for-personalized-nutrition-and-healthspan-optimization-301620452.html.
19 'Supplier Responsibility', Apple (Australia), accessed 16 July 2023, apple.com/au/supplier-responsibility.
20 'Timeline: Starbucks History of LGBTQIA2+ Inclusion', Starbucks, 26 February 2016, stories.starbucks.com/stories/2016/howard-schultz-dream-fulfilled-starbucks-to-open-in-italy.
21 Elon Musk, 'The Secret Tesla Motors Master Plan (Just Between You and Me)', Tesla Australia, 2 August 2006, tesla.com/en_AU/blog/secret-tesla-motors-master-plan-just-between-you-and-me.
22 Elon Musk, 'Master Plan, Part Deux', Tesla Australia, 20 July 2016, tesla.com/en_AU/blog/master-plan-part-deux.
23 J. Greathouse, '5 Time-Tested Success Tips From Amazon Founder Jeff Bezos', *Forbes*, 30 April 2013.

[24] At the same time, the letter likely energised the open-source software movement, which fundamentally disagreed with the letter's argument. Our digital world depends on open-source and free software, which is unlikely to change any time soon.

[25] At a very decent speed of 500 words per minute.

[26] Will Knight, 'OpenAI's CEO Says the Age of Giant AI Models Is Already Over', *Wired*, 17 April 2023.

[27] Jeremy Kahn, 'Who's Getting the Better Deal in Microsoft's $10 Billion Tie-up With ChatGPT Creator OpenAI?', *Fortune*, 25 January 2023.